Ammonia Fuel Cells

Ammonia Fuel Cells

Ibrahim Dincer

Osamah Siddiqui

ELSEVIER

Elsevier
Radarweg 29, PO Box 211, 1000 AE Amsterdam, Netherlands
The Boulevard, Langford Lane, Kidlington, Oxford OX5 1GB, United Kingdom
50 Hampshire Street, 5th Floor, Cambridge, MA 02139, United States

Library of Congress Cataloging-in-Publication Data
A catalog record for this book is available from the Library of Congress

British Library Cataloguing-in-Publication Data
A catalogue record for this book is available from the British Library

ISBN: 978-0-12-822825-8

For information on all Elsevier publications
visit our website at https://www.elsevier.com/books-and-journals

Publisher: Oliver Walter
Acquisitions Editor: Ruth Rhodes
Editorial Project Manager: Joanna Collett
Production Project Manager: Sreejith Viswanathan
Cover Designer: Alan Studholme

Typeset by SPi Global, India

Contents

Preface

During the past decade, ammonia has received increasing attention from various sectors and many people due to its excellent properties and carbon-free nature although it has been essentially very well known in the refrigeration industry for refrigeration applications. We have gone into an era where hydrogen is essential for many sectors as a fuel and as an energy carrier. Due to the considerable changes required for hydrogen energy systems and applications, ammonia appears to entail a vital position in the development of the hydrogen economy. Having numerous favorable properties and characteristics, ammonia is considered to be a promising candidate that can address several challenges faced by hydrogen usage as a fuel. Ammonia has, for many decades, been serving in various sectors, for example, fertilizers, working fluids, cleaning chemicals, refrigerants, and many more in addition to refrigeration systems as stated above. However, from the perspective of fuels, the necessity of shifting toward carbon-free fuels in this era has made ammonia stand out as a promising fuel that can aid in reducing the dependence on fossil fuel-based energy production. Also, ammonia is recognized as a clean solution as long as it is produced cleanly by renewables. In this regard, ammonia fuel cells play a key role toward the flourishment of ammonia energy. Entailing higher efficiencies than combustion engines, fuel cells along with other renewables are anticipated to be the upcoming revolution in power generation.

In this book, Chapter 1 first highlights the importance of implementing clean energy resources including the utilization of ammonia and hydrogen fuels and then discusses them in brief. Next, Chapter 2 covers the underlying fundamentals of fuel cell technologies as well as background information essential to understand the working methodology of different types of ammonia fuel cells. The necessary components required to constitute a fuel cell device are described and the physical as well as chemical phenomena occurring at each of these components is discussed. Also, the classification of fuel cells according to different classification categories is described where the categorization according to the type of electrolytes, fuel, and operating conditions is presented. Chapter 3 dwells on identifying different types of fuels that have been investigated for fuel cell applications. These range from hydrogen as the most commonly employed fuel to different types of alcohols as well as alkanes. Ammonia as a promising fuel is then described comprehensively presenting various advantages and favorable properties it entails. These provide sufficient information needed to proceed toward the details of ammonia fuel cells. Chapter 4 explicitly concerns the ammonia fuel cells where it comprehensively covers different types of ammonia fuel cells that have been developed. The performance of each type of ammonia fuel cell is linked to the type of electrolyte, electrochemical catalyst, and electrodes used. Ammonia fueled high-temperature solid oxide fuel cells with both proton-conducting and anion-conducting electrolytes is discussed. Furthermore, low-temperature direct ammonia alkaline fuel cells are covered in depth.

The development, materials, operation, system conditions, and catalysts of these ammonia fuel cells are presented. Moreover, Chapter 5 presents the analysis and modeling of fuel cells with specific focus on electrochemical interaction of ammonia molecules that occurs in direct ammonia fuel cells. The performance of these different types of ammonia fuel cells is also discussed. Chapter 6 describes various new integrated ammonia fuel cell-based systems that have been developed in the recent past and their performances are analyzed through overall energy and exergy efficiencies. Chapter 7 discusses novel ammonia fuel cell-based technologies as case studies where their performances are investigated at varying operating conditions and system parameters. Lastly, the book closes with Chapter 8 providing several recommendations for future development of ammonia fuel cells and depicting the development of ammonia-based energy technologies moving toward a new ammonia-based era.

Nomenclature

A	area (m^2)
C	charge (Coloumbs), concentration (mol/m^3)
cny	conductivity (S/m)
D	day angle (°), diffusion coefficient (m^2/s)
ex	specific exergy (kJ/kg)
E	potential (V)
$\dot{E}x$	exergy rate (kW)
F	Faraday's constant (96,500 C/mol)
g	specific Gibbs energy (J/mol)
G	Gibbs free energy (J)
$\bar{h}$	specific molar enthalpy (kJ/mol)
h	specific enthalpy (kJ/kg)
H	enthalpy (kJ)
$\dot{I}$	solar intensity
J	current density (A/m^2)
m	mass (kg)
$\dot{m}$	mass flow rate (kg/s)
M	molar concentration (mol/m^3)
n	number of electrons
$\dot{N}$	molar flow rate (mol/s)
P	pressure
$\dot{P}$	power (kW)
q	heat transfer (kJ)
$\dot{Q}$	heat transfer rate (kW)
R	ideal gas constant (J/mol K), Ohmic resistance (Ohm)
s	specific entropy (kJ/kg K)
$\bar{s}$	specific molar entropy (kJ/molK)
S	entropy (kJ/K)
ST	solar time (h)
T	temperature (°C)
v	specific volume (m^3/kg)
V	voltage (V)
w	specific work (kJ/kg)
$\dot{W}$	work rate (kW)

Greek letters

Ω	resistance (Ohm cm^2)
α	charge transfer coefficient

$\in$ half-cell potential (V)

ρ density (kg/m^3)

η efficiency

δ declination angle, diffusion layer thickness

τ resistivity

μ surface coverage

γ order of reaction

Subscripts

a   anode

act   activation

ads   adsorbed

ar   aerosol

b   backward

c   cathode

con   condenser

conc   concentration

conv   convection

d   density

dest   destroyed

diff   diffusion

dl   declination

E   electrode

elec   electric

ex   exchange, exit, exergy

f   formation, forward

ga   gas

gr   grid

gen   generation

in   incoming

L   limiting

lt   latitude

mb   beam

mix   mixture

nl   normal

Ohm   Ohmic

on   ozone

op   overpotential

or   open circuit

ox   oxidant

prod   products

PV   photovoltaic

reac   reactants

ref.   reference

rev	reversible
ry	Rayleigh
s	solar, surface
scs	solar constant
sen	sensible
sn	sun
T	temperature
TV	throttle valve
wr	water
zh	zenith

Acronyms

ABS	absorption cooling cycle
AEM	anion exchange membrane
AS	ammonia synthesis
ASR	ammonia synthesis reactor
DAFC	direct ammonia fuel cell
DEFC	direct ethanol fuel cell
DMFC	direct methanol fuel cell
DPFC	direct propanol fuel cell
COMP	compressor
CON	condenser
COP	coefficient of performance
Dl	day angle
EES	engineering equation solver
FC	fuel cell
GT	gas turbine
HF	heliostat field
HHV	higher heating value
HX	heat exchanger
KOH	potassium hydroxide
LHV	lower heating value
MCFC	molten carbonate fuel cell
MF	molar fraction
MEA	membrane electrode assembly
MSTR	multistage turbine
ORR	oxygen reduction reaction
PAFC	phosphoric acid fuel cell
PEM	proton exchange membrane
PSA	pressure swing adsorption
PTFE	polytetrafluoroethylyne
PV	photovoltaic

PVA	polyvinyl alcohol
RC	Rankine cycle
RO	reverse osmosis
SEP	separator
SDC	samaria-doped ceria
SOFC	solid oxide fuel cell
SRC	secondary Rankine cycle
ST	solar tower
WES	water electrolysis
YSZ	yittria-stabilized zirconia

Introduction

Since the industrial revolution, power generation has been playing a vital role in the advancement of any nation, which has, therefore, become a central element of any economy that drives the increase or decrease in the national production levels. A continuous and reliable generation of power is essential to attain a sustainable and stable economy as well as industrial sector. However, the increased dependence on fossil fuels for power generation in the recent decades has deteriorated the environment considerably. The change in the global primary energy demands since 2010 are depicted in Fig. 1.1. The energy demands increased by nearly 1.5% in the beginning of the previous decade. This rise in demands has decreased marginally in the beginning of this decade. Also, toward the middle of this decade, the primary energy demands had increased by a comparatively lower amount. Nevertheless, since 2016 the energy demands have risen sharply where the rise increased from 0.6% in 2016 to 2% in 2017. This has increased further to 2.9% in 2018. Hence, every year due to several factors, such as industrialization, urbanization and modern development, the specific energy demands tend to increase continuously. The change in the amount of increase has varied in the recent past, however, there has been a steady increase in the demands. Moreover, as the world swiftly moves toward a technology-enriched livelihood, the demands for energy are expected to rise rigorously. The energy demands are of particular interest due to their direct relation with environmental impacts. Currently, the global energy production heavily relies on fossil fuels and the increase in energy demands will directly affect the usage of fossils across the globe. This is attributable to various factors. First, the power generation sector that provides electricity to run any industry, corporate office, or transportation sector is dependent on fossil fuels across the globe. This implies that as technical and infrastructural development continues to proceed, the usage of carbon-rich and environmentally detrimental fuels continues to rise. In addition, in various sectors such as transportation, the current technology includes the usage of hydrocarbon fuels, which has led to significant environmental detriments across the globe.

Such alarming facts have led to serious concerns across the globe and in various countries attention is being paid toward reducing the dependence on as well as usage of fossil fuels. However, in several parts of the world, fossil fuel-based resources comprise the major energy sources and their dependence could not be reduced significantly until recently. Fig. 1.2 depicts the total usage of coal, oil, and natural gas across the globe since the year 2000. As can be observed from the figure, the usage of each of these carbon-rich energy resources has increased continuously.

Ammonia Fuel Cells. https://doi.org/10.1016/B978-0-12-822825-8.00001-3

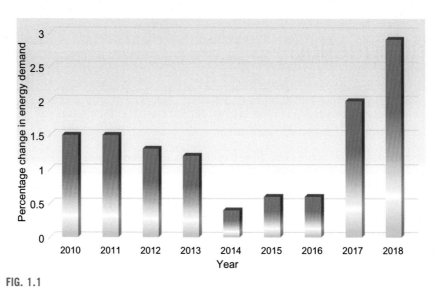

FIG. 1.1

Percentage change in global energy demand based on 2009.

Data from Ref. [1].

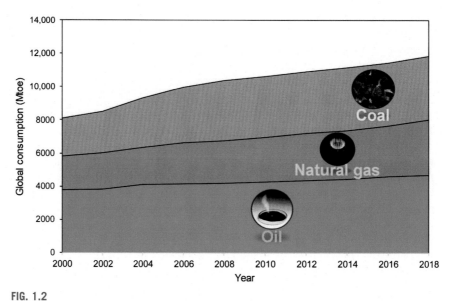

FIG. 1.2

Global consumption of coal, natural gas, and oil.

Data from Ref. [2].

Although various global efforts and agreements were made to reduce fossil usage as well as associated environmental impacts, the usage of such resources has increased steadily as shown in the figure. This is attributable to various factors such as increased energy demands. To meet the energy demands, these readily available resources are utilized extensively across the globe leading to considerable environmental damage. The usage of coal increased from nearly 2300 Mtoe in 2000 to nearly 3800 Mtoe in 2018. This denotes a significant increase of 65%, which shows the heavy dependence of various countries on fossil fuels across the globe. Similarly, the global usage of natural gas increased from approximately 2000 Mtoe in 2000 to nearly 3340 Mtoe in 2018. This also signifies an increase of 67% in the usage of the carbon-rich fossil fuel. In addition, oil is also used extensively in various sectors ranging from the industrial sector to the transportation sector. The usage of oil has also seen a steady increase in the recent years where an increase of nearly 25% is observed from 2000 to 2018. These are global estimates that constitute the usage by all countries. Some countries where fossil fuel-based energy resources are a source of income and aid in the development of the economy, entail a steady increase in the production as well as usage of carbon-rich fuels. However, these nations could essentially direct efforts toward the utilization of clean and renewable energy resources where applicable. For instance, some geographical locations receive high-intensity solar radiation across the year. Such locations should employ different types of solar-based power generation techniques. Similarly, some locations have high wind energy potential across the year that should be harvested. Also, other renewable energy resources including biomass, geothermal, and hydropower should also be considered where they are suitably applicable.

The recent breakdown of energy resource utilization across the globe is depicted in Fig. 1.3 where the global percentage of energy resources used for primary energy supply is presented. As depicted in the figure, majority of the energy supply was attained from carbon-rich resources. These included 27% coal, 22% natural gas, and 32% oil resources [3]. These three carbon-rich fossil fuels comprise nearly 80% of the total fossil fuel usage. The deployment of clean energy resources such as biofuels, nuclear, solar, etc., entails a minor portion of the overall usage. Although, after collective and collaborative efforts, the percentage usage of clean fuels has risen in the recent years, the present scenario still entails a heavy dependence on carbon-rich fossil fuels. The usage of biofuel and waste for power generation has gained pace and has been implemented in various countries across the globe. This entails the production of biofuels from waste that comprises useful fuels such as ethanol and biodiesel. Ethanol can be made from several plant-based raw materials that are generally referred to as plant biomass. Moreover, biogas is another type of biofuel that is produced from the anaerobic digestion of organic matter. Raw materials including manure, sewage, as well as agricultural waste can be deployed to produce biogas. This comprises mainly methane and carbon dioxide. Since methane has a sufficient value of combustion heat that is emitted when burnt with oxygen, it is used as a fuel for power generation. Although biogas also entails CO_2 emissions with its usage as a fuel, it is generally considered as a renewable energy resource owing to the

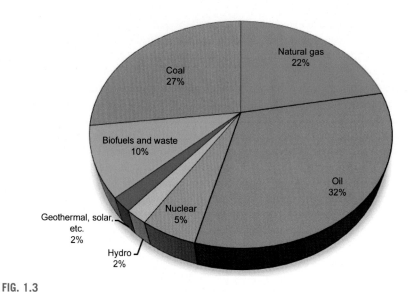

FIG. 1.3

Percentage of resources utilized for primary energy supply globally.

Data from Ref. [3].

carbon cycle. In this process, the CO_2 emissions resulting from the combustion of biofuel is recycled through its usage by plants during photosynthesis. Plants use CO_2 to synthesize glucose through the photosynthesis reaction that includes the production of glucose from water and carbon dioxide in the presence of sunlight. Moreover, nuclear power is also considered to be environmentally benign by various well-known organizations as it does not result in carbon emissions. However, there are other environmental as well as safety hazards associated with the usage of nuclear power that have raised concerns in various parts of the world. In addition, the nuclear power plant accidents and tragedies resulted in decreased attention toward nuclear-based power generation. These resources entail high carbon content and thus results in considerable amount of greenhouse gas (GHG) emissions.

The CO_2 emissions that resulted globally from energy-related fossil fuel usage in the past 20 years from 1998 to 2018 are depicted in Fig. 1.4. As depicted in the figure, there has been a significant increase in the CO_2 emissions in the past decade. In 1998, the global CO_2 emission from energy-related fossil fuel utilization was recorded to be 23.4 Gtonne. This increased significantly to 33.2 Gtonne in 2018 [4]. There exists a direct relation between the usage of fossil fuels for energy-associated activities and the increase in the CO_2 emissions, which have been identified to increase the global warming phenomenon. As can be observed from the figure, there is a trend of exponential rise in CO_2 emissions where the rate of increase in emissions with time is rising continuously. Every year the rate of change rises by a specific amount and the importance of reducing these emissions is evident from such observations. The exponential rise in CO_2 emissions can be linked to the increased usage of fossil

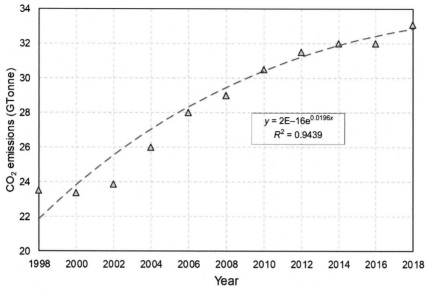

FIG. 1.4

Global carbon dioxide emissions arising from fossil fuel usage for energy-related activities which is also correlated in an exponential form.

Data from Ref. [4].

fuels as presented earlier. The rise in CO_2 emissions is particularly alarming owing to the greenhouse effect that it entails. Solar radiation entering the Earth's atmosphere is reflected by various objects and is trapped within the atmosphere due to the presence of GHGs such as CO_2. The trapped solar radiation leads to the trapping of thermal energy within the atmosphere of the Earth. This has been found to disturb the global temperatures where an increase in the average temperatures has been evidently proven in the recent years. This can have various detrimental effects on the ecosystem where the ice glaciers can start melting at alarming rates leading to a rise in sea levels. When sea levels rise, various dangers are posed toward countries situated near the oceans and other major water bodies. In addition to this, the weather cycles and associated cold as well as hot climate temperatures can also be affected by global warming phenomenon. The global warming phenomenon has raised major concerns about the sustainability and stability of the ecosystem as well as the future generations to come. Hence, there have been global concerns, efforts, and initiatives directed toward obtaining a solution to this increased fossil fuel dependence and the associated environmentally harmful emissions. Apart from carbon-based emissions, combustion of fossil fuels also results in the emission of nitrogen oxides (NO_x) and sulfur oxides (SO_x). These emissions have been proven to be detrimental to both human health and the environment.

Owing to these major concerns associated with fossil fuel usage, more environmentally benign energy production technologies, systems, and devices are being looked into with the objective of reducing the current environmental burden. Primary efforts in the recent past were directed toward solar, wind, and other environmentally benign power generation methods. However, the intermittency of these energy resources has hindered their widespread usage. Both solar- and wind-based power plants entail the disadvantage of not having a reliable and continuously stable input source of energy. Solar power plants require sufficient solar radiation input to operate and this depends heavily on the weather conditions in the specific area where the power plant is situated. Specifically, for solar thermal power plants (solar heliostat, parabolic trough, etc.), the availability of sufficient incoming solar radiation is necessary to heat the working fluid to the temperatures required for operating the plant. In addition to this, after sunset the solar-based power plants can only rely on the excess energy stored during the day. The conventional energy storage methods including thermal energy storage and batteries are capable of providing the required power for only a limited number of hours. Moreover, wind turbine-based power plants are also hindered by the intermittency of wind flow and velocities. At times, high-velocity winds occur that are sufficient to meet a high amount of load. Nevertheless, considerable fluctuations occur in wind velocities that make it difficult to predict the total power outputs of a given wind turbine power plant. Also, the wind velocities at times can be much lower than the minimum velocities required to operate the wind turbines. Furthermore, other renewable sources of energy such as hydro- and geothermal power plants are limited by geographical locations. Areas with high head water reservoirs or high flow rivers are suitable for hydropower plants, whereas areas that are deprived of these commodities cannot entail this power generation methodology satisfactorily. Furthermore, geothermal plants are also limited by geographical constraints. Locations where geothermal fluid is readily available and geothermal wells can be built are suited for such power plants. However, such resources are not readily available at all locations across the world. Hence, fuel cells are considered promising candidates that can play a vital role in the realization of an environmentally benign energy production infrastructure where the utilization of hydrogen or hydrogen entailing fuels provide a high-efficiency method for producing energy without harmful emissions. In addition, considering the transportation sector, there has been a significant rise in the GHG emissions arising from the use of fuels for transportation. Fig. 1.5 depicts the comparison of the CO_2 emissions arising from coal, oil, and natural gas globally in 1990 and 2017. The total emissions were estimated to increase from 20.5 Gtonnes CO_2 in 1990 to 32.6 Gtonnes CO_2 in 2017 [5]. This signifies an increase of nearly 60% during this time period. The CO_2 emissions arising from coal usage increased from nearly 8.3 Gtonnes in 1990 to 14.5 Gtonnes in 2017, which also constitutes an increase of nearly 60%. The CO_2 emissions are directly linked to the amount of usage of carbon-rich coal energy resources and this increase shows that their usage has risen considerably in the recent decades and the corresponding environmental impacts have also increased. Similarly, the CO_2 emissions arising from the usage of natural gas increased from

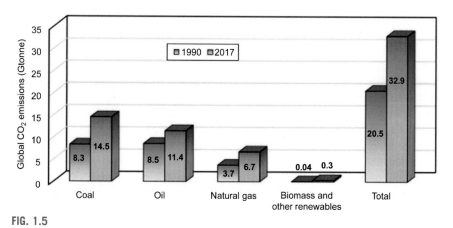

FIG. 1.5

Comparison of carbon dioxide emissions arising from different energy resources in 1990 and 2017.

Data from Ref. [5].

3.4 Gtonnes in 1990 to 6.7 Gtonnes in 2017. This indicates an increase of nearly 100% in the CO_2 emissions resulting from natural gas usage. This can be attributed to the increase in the implementation of natural gas-based industrial developments. Also, the total rise in CO_2 emissions in the recent decade can be attributed to the increased usage of fossil fuel vehicles along with industrialization during these years. Hence, several efforts have been exerted on the development of environmentally benign vehicles that are fueled by clean fuels such as hydrogen or ammonia or operate via battery power.

Clean transportation can play a key role in decreasing the environmental impacts that are being caused due to the usage of fossils. In some cities around the world, transportation sector has been identified as the major contributor toward the worsening of the air quality index. The emissions arising from automobile tailpipes include several types of harmful compounds that harm both the environment as well as human health. For instance, carbon monoxide is known to be detrimental to human health where it combines with the hemoglobin in the blood to form permanent compounds that can have adverse effects on human health. Also, particulate matter emissions that are emitted from the usage of different types of fossil fuels are also harmful for lungs and have been identified as one of the reasons for respiratory diseases. In some countries, such detrimental tailpipe emissions have forced the residents to wear pollution masks while outside. Moreover, the formation of smog is another harmful environmental detriment that is caused due to the increased usage of fossil fuels. When excessive smoke is released into the atmosphere in the presence of fog, a harmful mixture is formed in the atmosphere which is known as smog. This comprises several types of harmful compounds that can also lead to various respiratory diseases.

Hence, the utilization of automobiles operating with environmentally benign fuels is essential for environmental sustainability. Fuel cell vehicles that have been developed in the recent past primarily include the utilization of hydrogen fuel that is passed through a proton exchange membrane (PEM) fuel cell, which generates the required power needed by the vehicle. The central advantage of these vehicles lies in the fact that when hydrogen is used as the fuel, water (H_2O) is the only output emission. Thus providing a clean method for powering vehicles where no carbon entailing emissions arise. Also, PEM fuel cells entail higher efficiencies than conventional engines as discussed in the proceeding sections. This is also an advantage that is associated with the usage of fuel cell vehicles. In the recent past, several new fuel cell-based vehicles were introduced. The Hyundai Nexo, Honda Clarity, Toyota Mirai, F-Cell Mercedes-Benz are a few examples of fuel cell cars developed by automotive companies in the recent past. In addition to this, various locomotives and trains powered by fuel cells have also been developed in the recent years. The Alstom Coradia iLint is one of the latest commercially operated passenger train powered by fuel cell technology. Several other prototype trains, locomotives as well as buses were introduced recently across the globe [6]. Although efforts are being directed toward the development of hydrogen-based energy production for various sectors, several challenges hinder the flourishment of this technology. The high flammability of hydrogen associates it with a high safety risk while transportation as well as storage. Further, the low volumetric density of hydrogen gas necessitates compression to high pressures to attain an appropriate storage volume. Moreover, entailing an odorless characteristic, hydrogen is not easy to detect without proper hydrogen sensors. Thus, increasing the safety concerns owing to the high flammability risk in case of any leakages. Owing to such challenges, other hydrogen-based fuels have been investigated for fuel cell applications. Ammonia is a promising hydrogen-based fuel that entails the solutions to these challenges by having a low flammability range, higher density, higher boiling point, and easily detectable smell. Thus, ammonia has been investigated as a fuel for various power generation methods including fuel cells, internal combustion engines as well as ammonia gas turbines.

1.1 Historical background

The history of utilization of energy resources is known to have begun from wood as a source for obtaining thermal energy. In the earlier times, wood was burnt through sparks emitted from the rubbing of stones. The thermal energy obtained from this process was used for various purposes. An estimated timeline depicting the evolution of dominant and prevalent energy resources from wood to fuels with lower carbon to hydrogen ratios is depicted in Fig. 1.6. Until the mid-1700, the usage of wood as an energy resource was common that decreased considerably in the upcoming years. Through the exploration of coal as an effective fuel to generate thermal energy, its usage gained extensive popularity, especially in the industrial revolution. Specifically, in countries where considerable coal reserves were found, it was

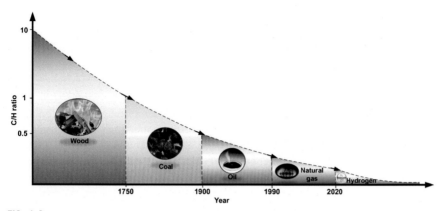

FIG. 1.6

Evolution of dominant energy resources during different times depicting the change toward low carbon to hydrogen ratio fuels.

adopted as the primary source of energy production across the nation. In addition, when coal powered locomotives were introduced, the usage of coal increased extensively in major parts of the world. Even today, some countries having massive coal reserves rely on the carbon intensive fuel for energy production. Next, the advent of oil reserves occurred where countries with significant amount of oil resources were observed to flourish massively. In few years, the usage hydrocarbon gases such as natural gas, propane, etc., became increasingly common in various countries owing to the discovery of new gas reserves. Also, the usage of gasoline powered cars and automobiles became a global commodity where transportation became heavily dependent on fossil fuels. Nearly all modes of transportation including road, air, and water transport mediums were manufactured to operate with hydrocarbon fuels. Although, the evolution of energy resources was dependent on numerous reasons, situations, and conditions, a general trend of decreasing carbon to hydrogen ratios can be observed. In the recent years, when the usage of hydrogen as a fuel gained pace, this trend further continued toward lower and lower usage of carbon-based fuels. Entailing zero carbon content, hydrogen fuel is considered as a promising alternative to fossil fuels that can overcome the massive environmental detriments, which the usage of fossils have caused in the recent past. The primary usage of hydrogen as a fuel comprises using electrochemical devices referred to as fuel cells.

Fuel cells work on the principle of the generation of electrical energy as a consequence of a series of electrochemical reactions. This phenomenon can be found in various aquatic animals including the eel, torpedo ray, etc., that are known to generate voltages of nearly 300 V [7]. The principle of utilizing fuel cells for electricity generation by providing the necessary electrolytes, ions, and electrodes is generally attributed to Sir William Grove, who had established the working principle satisfactorily in 1839.

Nevertheless, the electrochemical principles associated with the interaction of dissimilar metals producing electrical signals were introduced by Sir Luigi Galvani, Sir Humphrey Davy, and Sir Alessandro Volta in the early 19th century. These types of electrochemical cells were named the galvanic and voltaic cells. The electrochemical cell developed by Sir Humphrey Davy utilized the interaction between oxygen obtained via nitric acid and carbon atoms. This cell was particularly interesting in the sense that it could produce electrical energy via consumption of carbon entailing coal. However, the practicality required for useful applications was not achieved through this methodology. Nevertheless, these embodiments provided the initiative for developing new types of direct coal fuel cells, which entailed the working principle of using coal (readily available fuel at that time) to produce electrical energy directly without the usage of an intermediate heat engine. Moreover, the term "fuel cell" is reported to be coined by Mond and Langer [8]. Since these times, several efforts were directed toward proposing, developing, and investigating different types of fuel cells. Considerable progress was achieved as compared to the initial developments. These progressive efforts were not continuous throughout time and spans existed where people lost their interest in fuel cell technologies. The first time period where increased interest and efforts existed is generally stated between 1839 and 1890. During this period, the initial development of fuel cells as well as the introduction of electricity kept the keenness alive in fuel cell technology. Specifically, owing to the use of low-efficiency steam engines at those times, fuel cells were considered promising technologies that can aid in utilizing coal directly for power generation in a more efficient way. Efficiencies of more than 50% were expected through the coal-based fuel cell technologies. However, after several efforts, this idea was abandoned considering it to be impractical for useful applications.

The next time span after 1890s where there was considerable progress in fuel cell inventions existed between 1950 and 1960s. Significant efforts toward research and the invention of new and better fuel cell technologies started after the successful operation of the alkaline type fuel cell technology. These were satisfactorily utilized in spacecraft applications. The Apollo as well as Gemini programs developed by the United States raised positive expectations toward the development of more economically viable and technologically practical fuel cell energy systems. Furthermore, during this period several private as well as public initiatives and investments were made in fuel cell development. Many types of fuel cells popularly known today were studied and their advancement was initiated during this era. These include the popular PEM fuel cell, solid oxide fuel cell (SOFC), molten carbonate fuel cell (MCFC), and phosphoric acid fuel cell. Moreover, during this era the fuel cells were differentiated on the basis of the electrolyte mobility. The alkaline fuel cells, for instance, entailed electrolytes that could flow readily, however, the SOFC were associated with solid electrolytes. The choice of electrolyte was dependent on the application and usage. The initiatives undertaken during this era could not meet their expected outcomes and objectives for fuel cell development and application. This again led to a layback in the development of viable fuel cell technologies and the interests were withdrawn owing to the unfavorable outcomes. Nevertheless, the era filled with concerns regarding environmental sustainability led to the revival of the development of fuel cell technologies. Since early 1980s, continuous efforts have been exerted toward reducing the environmental burden caused due to the colossal usage of fossil fuels. Further, the human health detriments caused by the harmful emissions arising from fossil fuel usage as well as depleting fossil fuel reserves made the investigation and development of new environmentally benign energy resources inevitable. Several types of fuel cells were developed for commercial applications including PEM, AFC, SOFC, and MCFC in these years. Also, the fuel cell development in this era focused primarily on hydrogen fuel. However, several challenges hindering the flourishment of hydrogen fuel including high flammability, low volumetric density, and safety and economic disadvantages for transportation and storage have led to increased efforts in the development of alternative fuels. The detailed discussion about the challenges faced by hydrogen that hinder its current widespread usage are discussed in this book. Owing to those reasons, ammonia has been identified as a carbon-free fuel that can also be utilized for clean power production and it entails several advantageous properties and characteristics. Several types of ammonia-based power generation techniques exist. Energy can be obtained from ammonia fuel through conventional engines including compression or spark ignition systems. Nevertheless, the compression ratios required for ammonia combustion are different from diesel fuel. Further, due to specific flammability limits of ammonia, it has to be often blended with different combustion promoters that comprise conventional fuels. Moreover, ammonia gas turbines can also be employed for power generation where the gas-based power generation cycle comprising compression, combustion, and expansion processes can be implemented. In ammonia gas turbines, ammonia fuel is used instead of the conventional gas turbines that are fueled with natural gas as the fuel. Moreover, as ammonia is used in thermal engines through the process of combustion, it can also be used in electrochemical engines composed of different types of fuel cell technologies where electrochemical reaction of ammonia and oxygen occurs. This is the primary focus of this book. In 1960, Cairns et al. [9] studied the possibility of utilizing ammonia for generating electrical energy via fuel cells. They utilized a potassium hydroxide electrolyte, thus constituting an alkaline fuel cell type. Since then, several researchers and scientists have exerted efforts on the development of ammonia-based fuel cells for power generation that are described comprehensively in the proceeding chapters.

In this book, first, the underlying fundamentals of fuel cell technologies are described in detail as they provide the background information needed to understand the working methodology of different types of ammonia fuel cells. The necessary components required to constitute a fuel cell device are described and the physical as well as chemical phenomena occurring at each of these components are discussed. Further, the classification of fuel cells according to different classification categories is described where the categorization according to the type of electrolytes, fuel, and operating conditions is presented. Next, the book covers different types of fuels that have been investigated for fuel cell applications.

These range from hydrogen as the most commonly employed fuel to different types of alcohols as well as alkanes. Ammonia as a promising fuel is then described comprehensively presenting various advantages and favorable properties it entails. Further, ammonia fuel cells are covered comprehensively where development of different types of ammonia fuel cells are described. The performance of each type of ammonia fuel cell has been linked to the type of electrolyte, electrochemical catalyst, and electrodes used. Ammonia fueled high-temperature SOFCs with both proton-conducting and anion-conducting electrolytes is discussed. Also, low-temperature direct ammonia alkaline fuel cells are covered in depth. The development, materials, operation, and system conditions of these ammonia fuel cells are presented. Moreover, the analysis and modeling of fuel cells is presented with a specific focus on electrochemical interaction of ammonia molecules that occurs in direct ammonia fuel cells. The performance of these different types of ammonia fuel cells is also discussed. Several new integrated ammonia fuel cell-based systems that have been developed in the recent past are presented and their performances are assessed through overall energy and exergy efficiencies. Lastly, novel ammonia fuel cell-based technologies are presented as case studies and their performances are investigated at varying operating conditions and system parameters. The book closes with several recommendations for future development of ammonia fuel cells identifying the key challenges faced by these technologies.

1.2 Closing remarks

In this chapter, the importance of reducing the dependency on fossil fuel-based energy production is highlighted. The rising energy demands across the globe are shown through recent statistics covering. Furthermore, the corresponding rise in CO_2 emissions is explained showing the trend in the exponential rise in these emissions in the recent years. Also, a comparison of CO_2 emissions from carbon-rich coal, natural gas, oil, and biomass energy resources is made comparing recent emission data with recent decades. The statistics are alarming and show that immediate attention is needed toward the reduction of fossil fuel usage and rise of renewable energy utilization. Moreover, a historical perspective of energy resources is presented showing the trend toward the utilization of low carbon content energy resources, where the initial usage of wood as an energy fuel was replaced with coal when coal mining was introduced. Further, the advent of natural gas energy reserves led to the shift toward natural gas-based power generation in various parts of the world. Further, after oil was introduced along with its various by-products for utilization in various industries, its usage gained momentum. Next, the increased attention toward low-carbon fuels led to the usage of carbon-free hydrogen fuel in recent years. The historical background of fuel cell technologies is also presented providing an overview of the technological development and advancements that occurred since the 19th century.

Fundamentals

In this chapter, the fundamentals of fuel cells are presented and comprehensively elucidated. The concept of hydrogen economy is first discussed followed by the underlying operating phenomena of fuel cells. The similitude of fuel cells to other electrochemical devices is presented and the essential components constituting a fuel cell device are described. Furthermore, the chemical and physical phenomena that occur within a fuel cell during the electrochemical interactions are explained. The classification of different types of fuel cells according to the type of electrolyte, electrode, operating temperature, etc., is also presented.

2.1 Hydrogen economy

The hydrogen economy entails the concept of utilizing hydrogen as the prime fuel for various applications. Some of these applications include electricity generation, transportation, metal refining, fertilizer production, synthetic fuel production, etc. The ecosystem based on hydrogen comprises clean electricity generation from solar, wind, or nuclear power. The electricity produced from such environmentally benign energy resources is supplied to the grid and the excess energy available in the form of solar or wind energy is stored in the form of hydrogen. Hydrogen is synthesized via water electrolysis that does not entail carbon emissions, which are associated with steam methane reforming-based hydrogen production. The hydrogen synthesized from these clean energy resources is used as a fuel to power automobiles as well as other transportation vehicles. Further, the produced hydrogen is also used to synthesize synthetic fuels such as ammonia. Also, other applications include industrial usage such as refining of metals. In addition to this, during periods of low energy availability, the produced hydrogen can be used as a fuel to generate electricity through the usage of fuel cells. Moreover, to utilize hydrogen as a fuel in transportation vehicles, fuel cell technologies are required. Hence, fuel cells entail high significance in the hydrogen economy. Fuel cells entail some resemblance to batteries. Consider a non-rechargeable battery as depicted in Fig. 2.1. In such chemical to electrical energy conversion devices, the reactant fuel as well as the required reactant oxidants are stored within an enclosed containment. Owing to this principle, they are sometimes known as onboard storage devices. Batteries are thus energy storage devices such that they store chemical energy in the form of the fuel and oxidants.

Ammonia Fuel Cells. https://doi.org/10.1016/B978-0-12-822825-8.00002-5

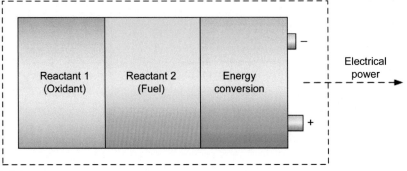

FIG. 2.1

Schematic representation of a battery with enclosed reactant fuel and oxidant.

When electrical energy generation is required, this stored chemical energy is converted into electrical energy via electrochemical interactions between the stored reactant fuel and the oxidants. Hence, each battery has a specific amount of output electrical energy that it can supply. This amount is directly dependent on the amount of reactant fuel and oxidant entailed in the enclosed battery containment. In the case of a non-rechargeable battery, the maximum output electrical energy is thus limited by the enclosed reactant amount. Once they completely react with each other, no more chemical energy to electrical energy conversion is possible. The lifespan of a battery power reserve is thus limited, according to the chemical reactants satisfactorily enclosed. In addition to this, as all the reactants required for the electrochemical conversion are enclosed together, the electrochemical interactions begin even before power is extracted from the battery for useful purposes. Although these interactions occur at a slow rate, if batteries are unused for long periods of time, these interactions may cause a large loss of useful output potential. This is another limitation of storing energy in the form of batteries. Furthermore, the electrode employed in a battery device is also consumed during the electrochemical conversion. It is thus sometimes stated that the battery lifetime is a function of the electrode lifetime. The higher the electrode lifetime, the longer will be the lifetime of the battery unit.

However, a fuel cell (so-called: an electrochemical device) differs from a battery primarily with the principle that the reactant fuel as well as oxidant are not stored within the cell and are continuously supplied from an external source or fuel reservoir. A simplified schematic representation of this is depicted in Fig. 2.2. A fuel reservoir that exists outside the fuel cell stores the fuel and supplies according to the required power output needed from the cell. Also, the oxidant needed for the electrochemical reaction is obtained externally usually from ambient air or an oxygen reservoir if necessary. Moreover, as the electrochemical interactions take place between the fuel and the oxidant, the reaction product is formed on an electrode that needs to be extracted from the cell continuously. Also, the exothermic reactions produce heat, which also needs to be extracted from the fuel cell to prevent overheating of the cell that can deteriorate the performances significantly. Therefore, unlike

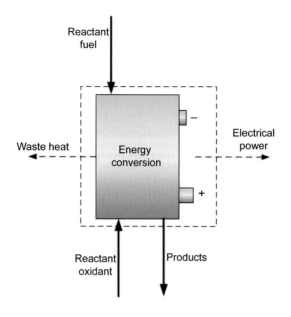

FIG. 2.2

Simplified schematic representation of a fuel cell operating with external fuel and oxidant inputs.

battery systems, fuel cells can operate continuously as long as they receive the fuel and oxidant inputs. In comparison, battery systems can only operate in conjunction with the reactant amount enclosed within the containment. Due to this, fuel cells are considered independent of lifetime spans as far as the fuel is concerned. They can operate continuously as long as a fuel input feed is provided. However, other fuel cell components including the catalyst layers, membranes, electrodes, etc., have lifetimes, which are sufficiently long as compared to a battery system. Moreover, as the fuel cell does not store the reactants, there is no loss of energy in the case of non-utilization of the device for long periods as for the case of batteries. In fuel cell systems, the fuel is safely stored in cylinders or containers that do not allow any chemical interaction between the fuel and any other external chemical compound. The fuel is directly passed from the storage tanks to the fuel cell device. Also, the electrodes in fuel cell devices do not generally participate in the electrochemical reactions and are thus not consumed during the process. This is not the case for battery systems, where electrodes play a key role in the electrochemical reactions occurring within the cell.

Nevertheless, there are several similarities between fuel cells and batteries. First, both entail the principle of converting chemical energy directly into electrical energy. Furthermore, both utilize electrochemical interactions for producing electrical energy. Hence, both are electrochemical energy conversion devices. Also, both entail the usage of electrolytes and electrodes (anodes and cathodes). The term "battery," however, can be associated to Sir Benjamin Franklin, who has utilized this term to describe a collection of capacitors that were also known as Leyden jars.

2.2 Fuel cells and heat engines

As discussed earlier, fuel cells and heat engine-based electrical generators also entail similarities in their overall energy conversion principles. Both utilize the inherent chemical energy of fuel to generate electrical energy. However, in a heat engine-based electricity generation system, the chemical energy of the input fuel is firstly converted into thermal energy. The input fuel is combusted in the presence of an oxidant and the resulting combustion chemical reaction generates heat owing to its exothermic nature. The heat or thermal energy that is generated is then utilized further to produce mechanical energy. The resulting mechanical energy is further utilized to generate electrical energy via an electric generator that entails current carrying coils and thus magnetic flux, which generates electrical energy in the presence of rotational kinetic or mechanical energy of the shaft. Thus, a heat engine-based electricity generation method entails a series of energy conversion steps that start with the conversion of chemical energy entailed in a fuel into thermal energy, which is further converted into mechanical energy that is converted finally into electrical energy. A simple schematic representation of heat engine is depicted in Fig. 2.3. Most importantly, every energy conversion step involves losses as each type of energy conversion includes irreversibilities and exergy destructions. Not all the chemical energy entailed in the reactants can be converted into thermal energy and not all the thermal energy produced by fuel combustion can be converted into mechanical energy. Furthermore, the mechanical energy of the rotating shaft cannot be completely converted into electrical energy. On the other hand, a fuel cell entails direct conversion of chemical energy into electrical energy, which allows the avoiding of several energy conversion steps and thus resulting in higher energy efficiencies. In addition, heat engine-based electrical energy generation includes the limits posed by the Carnot efficiency. Since no practical heat engine can have an efficiency of more than the Carnot engine, the Carnot efficiency that depends on the high and low temperatures in the system define an upper limit for the efficiency for any given heat engine. The high temperature of the heat engine system that directly affects the Carnot efficiency is a function of the fuel combustion reaction and thus is also limited. These limitations can be due to several factors such as the heat losses, heat absorptions by side processes and reactions as well as conversion ratios. Also, heat engines need to be equipped with several moving parts and components. These parts always involve frictional losses as well as performance degradation with time due to wear. These factors also contribute to decrease in the efficiencies. In addition to efficiency limitations, heat engines are also associated with environmental detriments that have raised major concerns across the globe in the recent

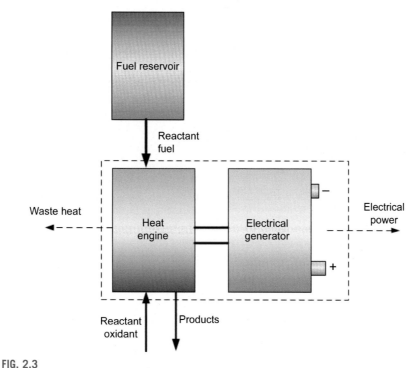

FIG. 2.3

Simplified schematic representation of a heat engine-based electricity generation system.

past. Conventional electricity generation power plants involve the usage of heat engines, which are mainly powered by fossil fuel resources such as coal or natural gas at present. When such fossil fuel resources are utilized, they result in large amounts of GHG emissions as well as other harmful emissions that are detrimental to both the environment and the human health.

Considering the above-discussed issues, fuel cells are seemed to be promising electricity generation devices as compared to heat engines. This is due to the possibility of generating electrical energy at a higher efficiency than a heat engine. Also, the energy conversion is a single step process rather than several conversion steps as in the case of a heat engine and thus involved lower amount of irreversibilities and losses. Further, unlike heat engines, fuel cells do not include several moving parts that involve regular deterioration and thus require maintenance, resulting in lower maintenance costs. Nevertheless, fuel cells are also limited by several factors. These include deterioration of catalyst layers, the catalyst layers that are developed to aid in the electrochemical reaction also degrade with time. The time for degradation depends on the type of catalysts, membranes, and fuels utilized. Also, other components including electrodes and gas diffusion layers can also get damaged overtime.

2.3 Fuel cell working principles

A given fuel cell device comprises three essential building blocks that are necessary for the operation of electrochemical interactions. First, the anode or the fuel electrode where the fuel is input to the cell and the anodic electrochemical reactions occur in the presence of catalysts. Second, the cathode or the oxidant electrode where the oxidant reactants are input to the cell and they react electrochemically according to the cathodic electrochemical reactions of the cell. Third, the electrolyte that is generally situated between the anode and cathode. The electrolyte is also a key component necessary for the transfer of ions within the cell. A schematic representation of these fuel cell components and the corresponding functions for a proton exchange membrane (PEM) hydrogen fuel cell are depicted in Fig. 2.4. As depicted in the figure, the fuel, that is, hydrogen in this case enters at the anode side of the cell. The anode or the fuel electrode allows the passage of hydrogen gas to reach the catalyst layers where they react electrochemically and emit electrons.

The anodic electrochemical reaction in this case can be represented as

$$H_2 \rightarrow 2H^+ + 2e \qquad (2.1)$$

The electrons emitted as a result of this half-cell electrochemical reaction generate the current in the circuit that can include an external load through which electron flow is generated. As the anodic electrochemical reaction keeps emitting electrons, the corresponding cathodic reaction associated with accepting electrons is denoted by

$$\frac{1}{2}O_2 + 2H^+ + 2e \rightarrow H_2O \qquad (2.2)$$

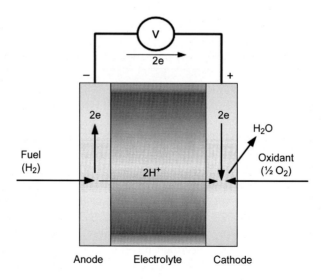

FIG. 2.4

Schematic representation of a hydrogen PEM fuel cell.

where the positively charged H^+ ions are transferred to the cathodic side of the cell through the membrane electrolyte. In PEM fuel cells, the electrolyte comprises a membrane that allows the passage of positively charged H^+ ions, which are often referred to as protons. The oxygen needed at the cathode to complete the cathodic electrochemical reaction is supplied from an external source. If ambient air is utilized as the oxygen source, it also entails nitrogen that exits the cathodic compartment unreacted. As can be observed from the cathodic reaction, the final product of the overall reaction is water (H_2O). Thus, the series of ion, electron, and mass transfer steps keep producing electrical energy that can be utilized for useful purposes. The overall reaction for a hydrogen fuel cell is expressed as

$$H_2 + \frac{1}{2}O_2 \rightarrow H_2O \qquad (2.3)$$

In hydrogen fuel cells, as water is formed continuously at the cathode, it needs to be removed from the cells as the accumulation of water causes flooding in the cell. High accumulation of water molecules hinders the mass transport and thus affects the overall electrochemical activity. In addition, the overall reaction being exothermic results in the generation of heat. Hence, continuous heat removal from the fuel cell is necessary to prevent damage of cell components.

The working principles of ammonia fuel cells are similar to hydrogen fuel cells entailing electrode reactions as well as membrane electrolytes. However, alkaline electrolyte-based ammonia fuel cells entail several differences as compared to hydrogen fueled PEM fuel cells. A schematic representation of an alkaline electrolyte-based direct ammonia fuel cells (DAFCs) is shown in Fig. 2.5 illustrating the reactants and products at both the anode and the cathode.

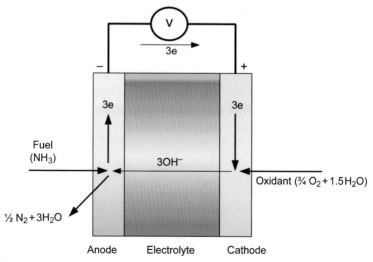

FIG. 2.5

Schematic representation of a direct ammonia fuel cell.

The fuel input feed in a DAFC comprises a direct inlet of ammonia (NH_3). In alkaline-based DAFC, this input fuel reacts electrochemically at the anode in the presence of catalyst with negatively charged hydroxyl ions. This anodic reaction is represented by

$$NH_3 + 3OH^- \rightarrow \frac{1}{2}N_2 + 3H_2O + 3e \tag{2.4}$$

The above anodic electrochemical reaction emits electrons that generate a flow of electrons through any external load and allow useful power utilization. The corresponding cathodic electrochemical reaction that accepts electrons and completes the overall reaction is represented as

$$\frac{3}{4}O_2 + 1.5H_2O + 3e \rightarrow 3OH^- \tag{2.5}$$

The necessary cathodic reactants of oxygen (O_2) and water (H_2O) molecules need to be provided at the cathode, which also react in the presence of an electrochemical catalyst to form hydroxyl (OH^-) ions. These hydroxyl ions formed at the cathode migrate to the anodic side of the fuel cell through an alkaline electrolyte that allows the passage of only negatively charged anions. Hence, in this way, the electrochemical reactions keep occurring at either electrodes with fuel and oxidant inputs and useful power can be continuously extracted from the cell. The overall fuel cell reaction for an alkaline electrolyte-based DAFC can be represented as

$$NH_3 + \frac{3}{4}O_2 \rightarrow \frac{1}{2}N_2 + 1.5H_2O \tag{2.6}$$

There are other types of DAFC that have been developed in the recent past, which will be discussed in detail in the upcoming sections of this book. Although the working principles of DAFC are satisfactorily well established, their performance is considerably low as compared to hydrogen fuel cells. This is primarily attributed to the insufficient electrooxidation of ammonia molecules. The type of catalyst activity observed in the case of hydrogen oxidation for fuel cells has not been achieved yet for the case of ammonia oxidation. This remains one of the key factors that prohibit the development of high-performance DAFC.

2.4 Chemical and physical phenomena in fuel cells

The working principles described for hydrogen or ammonia fuel cells in the preceding section represent the overall processes. However, there are several other phenomena that occur during the operation of fuel cells and affect the electrochemical behavior and thus the performance. These can be mainly classified into chemical and physical phenomena. The underlying chemical phenomenon includes processes of dissociation of hydrogen molecules to hydrogen atoms and the subsequent adsorption process that is followed finally by the oxidation process. The physical phenomenon includes mass and species transport processes of the reactants as well as products that travel to and from the reaction sites to the exit and bulk concentration sides. The transport phenomenon occurring within the cell affects various parameters including voltage losses that occur within the cell as well as the transfer of heat. For instance, consider a fuel cell comprising a liquid electrolyte. As the fuel enters the cell, it comprises a mixture of the fuel as well as some minor amounts of water vapor, hydrocarbons, or other carbon-based compounds depending on the source of the fuel. The fuel mixture or the oxidant mixture reaches the surface of the electrode by means of convection. The electrode is generally made of porous material to allow the passage of reactant or oxidant molecules and by means of diffusion, these gases arrive at the adjacent interface comprising the electrolyte and reactant gas contact. Next, in the case of a liquid electrolyte, dissolving of reactant molecules occurs at this interface. This interface comprises both the liquid electrolyte and the gaseous reactant and is thus often referred to as two-phase interface. Furthermore, after dissolving in the electrolyte, diffusion of these dissolved species occurs from the two-phase interface to the surface of the electrode. During these processes, some undesired reactions may also occur. These include the reactions between the electrolyte and any impurities that were entailed in the reactant mixture. Also, corrosion of the electrode may occur due to contact with any unwanted oxidizing species. However, the next major process comprises adsorption. Those species that react electrochemically are first adsorbed on the surface of the electrode. The migration of adsorbed species on the surface of the electrode occurs by diffusion. Next, at the surface of the electrode that is in contact with the liquid electrolyte, the electrochemical reactions take place due to which positive or negative ions as well as electrons are generated. This interface is generally referred to as the three-phase boundary. As the ions accumulate at the surface of the electrode, diffusion of these charged species occurs across the surface. Further, the products of the electrochemical reaction are desorbed from the surface of the electrode. In the electrochemical cell, due to the formation of opposite charges across the two electrodes, an electric field is created that aids in the transfer of electrons as well ions. The electrochemical reaction products that are desorbed, diffuse through the electrolyte to reach the two-phase interface from where they finally exit the cell.

A schematic representation of these processes is depicted in Fig. 2.6. The gaseous fuel (considered H_2 in this case) enters the cell from the bottom and reaches the electrolyte and fuel interface. Further, the molecules dissolved at this interface diffuse through the electrolyte and reach the surface of the electrode and the adsorption phenomenon proceeds. After the process of adsorption, the electrochemical interaction occurs resulting in the formation of ions and electrons.

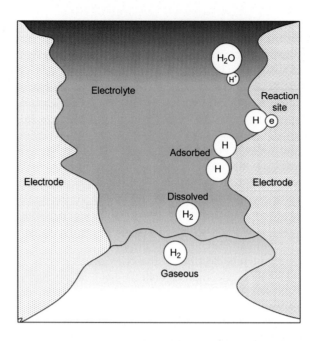

FIG. 2.6

Different chemical and physical phenomena occurring in a liquid electrolyte fuel cell.

2.5 Electrodes and electrolytes in fuel cells

As can be observed from the fuel cell processes and phenomena discussed earlier, electrodes as well as electrolyte are central fuel cell components that are needed for the electrochemical interactions to occur and produce useful electrical energy. There are three ways in which the utilization of electrodes in fuel cells can be described. First, electrodes provide the required reaction sites necessary for the dissociation, adsorption, and thus the electrochemical reaction to occur. Second, electrodes also provide appropriate flow channels for the fuel cell reactants as well as products. Especially, in the case of fuel cell stacks, electrodes with suitable flow channels are designed and used that allow smooth passage of reactants over each cell. Third, electrodes also act as electron collectors. When electrons are emitted during an electrochemical reaction, they first accumulate at the electrode that creates an electric potential between two electrodes. This acts as the driving force to move the electrons through an external circuit where the load is situated. One important feature of electrodes is their porous structure. As a solid structure is needed for providing adsorption and reaction sites, and an empty passage space is needed for the flow of reactants and products, a porous structure is desired for electrodes. Furthermore, as the electrodes are subjected to various types of reactants as well as oxidants, their chemical stability is essential for appropriate fuel cell operation. In case of any undesired side reactions between the electrode and reactants or products, the performance of the fuel cell deteriorates considerably. In addition, good mechanical strength, especially high compressive strength, is desired for fuel cell electrodes. The electrodes with porous structures are generally developed in layered configurations. The first layer comprises the porous structure that allows the passage of reactants and collects electrons (in conventional fuel cells these are known as gas diffusion layers). The second layer comprises a catalyst layer along with the interface of the electrolyte. Conventionally, platinum is the most commonly employed metal that is utilized for fuel cell electrodes. However, owing to its high cost, several research studies are being conducted to develop catalysts comprising nonprecious metals such as nickel or iron. In the case of ammonia fuel cells, platinum catalysts are poisoned during the dissociation and adsorption processes. This affects the performance of DAFCs adversely resulting in low open-circuit voltages, power outputs, and current densities. Due to this reason, several studies in the recent past have focused on developing new types of electro-catalysts that are compatible with electrochemical interaction of ammonia molecules. The different types of catalysts developed for ammonia fuel cells and their respective performances are described in Chapter 4.

The electrolyte is also a key fuel cell component that provides the medium required for the ions to transfer from one electrode to the other electrode. The functions of the electrolyte can also be summarized in three ways. First, electrolyte is the ion conducting medium that allows the ionic species formed during the electrochemical reactions to travel from the electrode where they are formed to the counter electrode to complete the half-cell electrochemical reactions. Second, the electrolyte also acts as an electric insulator that prevents the cell from short circuiting. A short circuit occurs in a fuel cell if the anode and cathode electrodes come into contact. In this case, the electrons are prohibited from traveling through any external load and the anode–cathode contact short circuits the cell. Thus, the electrolyte that is generally sandwiched between the two electrodes prevents any electrical contact and thus any chances of cell short circuit. Third, the electrolyte also separates the two compartments of the fuel cell that comprise the fuel input side and the oxidant input side. The electrolytes are made from materials that do not allow the cross flow of the fuel or oxidant from either side. Cross flow of reactants, also known as crossover, can adversely affect the fuel cell performance. Also, if the fuel utilized is combustible and crosses over to the oxidant side, there can be safety hazards. Hence, a suitable fuel cell electrolyte is required to entail several properties that make it compatible for usage. Insoluble and nonporous properties toward reactants are key characteristics. Further, high ionic conductivity is desirable for efficient flow of ions through the electrolyte, which is essential to achieve high fuel cell performances.

2.6 Performance of fuel cells

A complete fuel cell reaction comprises two half-cell electrochemical reactions that occur at either electrode. At one electrode (anode), the electrochemical reaction emits electrons (anodic reaction) and at the other electrode (cathode), the electrochemical reaction accepts electrons, thus, completing a full circuit. Since each electrode is accompanied by an electrochemical reaction, it entails an electric potential that is commonly referred to as the half-cell potential. The anodic electrochemical reaction gives rise to an anodic electric potential and the cathodic electrochemical reaction results in a cathodic electric potential. The overall fuel cell potential is thus determined from the difference between these two half-cell potentials. The electrons are generated at the anode and flow toward the cathode through an external circuit, hence the electron flow is from the anode to the cathode. However, according to the conventional current direction terminology, the current would be stated to be flowing in the opposite direction of electron flow (cathode to anode). The overall theoretical fuel cell potential for a hydrogen fuel cell considering Eq. (2.1) as the anodic reaction and Eq. (2.2) as the cathodic reaction is 1.229 V. For an ammonia fuel cell, the theoretical potential is 1.17 V considering Eqs. (2.4) and (2.5) as the anodic and cathodic electrochemical reactions, respectively. These potentials are theoretical and are obtained if the fuel is considered completely pure and at ambient conditions of 1 atm pressure and 25°C. Furthermore, these potentials are generally known as reversible fuel cell potentials as they are evaluated for thermodynamically reversible operation. This is discussed in detail in Chapter 5.

In actual fuel cell operation, several irreversibilities and losses occur inevitably. These losses occur primarily in the form of potential losses and owing to these irreversibilities, the working cell potential reduces substantially. For instance, for a hydrogen fuel cell, the working cell potential might drop to nearly 0.7 V if the optimal power density ranges are targeted in practical applications of power outputs. Also, in the case of the ammonia fuel cells, several undesired losses can drop the open-circuit voltage to nearly 0.3 V. Furthermore, a well-established phenomenon for fuel cells is the drop in cell voltage with an increase in the current densities. As the current drawn from the cell is increased, the cell voltage is observed to decrease due to several voltage losses. In fuel cells, the losses in voltage due to an increase in current densities is generally known as overpotential or overvoltage. Also, the physical or chemical phenomenon causing this overpotential is known as polarization. Generally, the cathodic electrochemical reaction entailing the reduction of oxygen molecules is associated with the highest loss in voltage. This can be attributed to the comparatively slower rate of the electrochemical oxygen reduction reaction (ORR). When a higher current is drawn from the cell, the number electrochemical reactions occurring within the cell also increase simultaneously to meet the higher electron flow required. However, due to rate-limiting factors such as hindrances in the mass transfer, the actual cell operating voltage reduces. Thus, one of the current challenges in fuel cell technologies is to overcome the slow

ORR reaction rates. Moreover, in hydrogen fuel cells, the anodic electrochemical reaction is satisfactorily fast and thus results in substantially lower voltage losses as compared to the cathodic counterpart. The voltage loss due to anodic electrochemical reaction in hydrogen fuel cells is thus negligibly small during actual fuel cell operation as compared to the voltage losses occurring due to ORR. However, in the case of ammonia fuel cells, unavailability of appropriate catalysts for the electrochemical oxidation of ammonia inhibits satisfactory anodic performance. The conventional platinum catalysts utilized in hydrogen fuel cells have been found to deteriorate considerably when exposed to ammonia, thus, adversely affecting the anodic half-cell performance. In the recent past, several types of alternative catalysts have been investigated to overcome this challenge faced by DAFCs. Specifically, in the case of low-temperature DAFCs, ammonia does not dissociate into its constituents of hydrogen and nitrogen before the electrochemical reaction. Thus, all ammonia molecules entering the anodic compartment are required to oxidize electrochemically. However, for high-temperature DAFCs such as ammonia fed solid oxide fuel cells (SOFCs), ammonia molecules generally dissociate into nitrogen and hydrogen. The hydrogen molecules formed because of this dissociation then react electrochemically at satisfactory rates in the presence of conventional platinum catalysts.

Furthermore, any electrical circuit is associated with Ohmic losses. Similarly, in an operating fuel cell where the electrochemical reactions lead to a flow of electrons that passes through an external load, Ohmic voltage losses occur. Among all fuel cell components, the electrolyte is associated with the highest Ohmic loss. The electrolyte resistance indicates the hindrance to the flow of ions. The higher the electrolyte resistance, the higher will be the amount of irreversibilities that occur during the ionic flow. In addition to this, when high currents are drawn from the cell, the electrochemical activity as well as mass transport are aggravated. The maximum current that can be drawn from the fuel cell depends on these mass transport limitations. Beyond a certain point, the mass transport phenomenon is significantly limited such that voltage drops exponentially and no more practical useful power outputs can be attained.

The phenomena that lead to these voltage losses are referred to as polarization losses. The first type of polarization loss is known as activation polarization. This refers to the voltage loss that occurs due to the reaction kinetics of the half-cell electrochemical reactions. These type of irreversibilities are much higher in the low current density region. The second type of polarization is known as Ohmic polarization. This refers to the voltage losses that occur due to the Ohmic resistances of the fuel cell components including the electrolyte, electrode, etc. The third type of polarization is known as concentration polarization. This refers to the voltage losses that occur due to mass transport limitations. Especially, in the regions of high output current, this type of polarization significantly affects the fuel cell performance. Hence, in order to improve the fuel cell technology, several studies are being conducted to develop new types of catalysts that can provide faster reaction kinetics for electrochemical reactions. Also, low-thickness electrolytes as well as electrodes are being developed with the objective of reducing their Ohmic resistances. Further, new types of materials are being developed and investigated to be utilized in fuel cell applications that can improve the performances by lowering the irreversibilities. Moreover, new types of materials entailing higher diffusion coefficients are also being developed.

Another important factor about the usage of fuel cells to obtain useful power output for practical applications is the theoretical voltage of a single cell. For hydrogen fuel cells, the theoretical cell voltage is 1.229 V and for an ammonia fuel cell, the theoretical voltage is 1.17 V. Hence, it is not feasible to achieve useful power outputs that can be utilized for practical applications through the utilization of a single cell. Therefore, fuel cells are arranged in stack arrangements. In a stack arrangement, several single cells are connected in series that provide a higher overall fuel cell stack voltage as well as power output. The fuel cell stacks are arranged in a way that connects the anode of one cell to the cathode of the next adjacent cell. This is achieved through the utilization of bipolar plates. Bipolar plates have become essential fuel cell components in the recent past that allow several cells to be stacked together in a compact way. Also, these plates are embedded with suitable flow channels that allow appropriate reactant flow over the electrodes. The flow channels and the appropriate stack arrangement ensures the reactant fuel flows over the anode side of each single cell and the oxidant flows over the cathode side of each cell. Therefore, fuel cell development in the recent past has also comprises new designs of bipolar plates, new configurations of fuel and oxidant flow, new designs of gas flow channels, etc. These elements also play a vital role in determining the performance of fuel cell stacks.

2.7 Advantages and applications of fuel cells

Unlike heat engines, fuel cells entail a single energy conversion step comprising chemical to electrical energy conversion. This provides fuel cell technologies with a significant advantage in terms of efficiencies. Even in practical applications and usages, fuel cells can operate at much higher efficiencies than heat engines. For instance, fuel cell applications in space shuttles operating with hydrogen fuel and pure oxygen as the oxidant have been reported to provide clean energy with efficiencies of more than 70%. In addition, along with this highly efficient energy conversion, only water is emitted, which can be used for useful purposes. Especially, in space applications, where all resources including fuel and water need to be utilized very effectively, fuel cells provide a promising methodology to produce power from hydrogen and oxygen to obtain water, a useful commodity that is used for drinking by the space crew.

Another key factor entailing the high efficiency of fuel cells is the advantageous variation in efficiencies from full to partial electrical loads. When the electrical output load provided by the fuel cell is decreased from the normal design operation load, the efficiency is expected to increase. However, this is not the case with heat engines where under partial load conditions, the efficiency due owing to higher irreversibilities. When a fuel cell is operated at lower or partial electrical loads, the cell potential increases as a result of a drop in the current density drawn from the cell. Thus, as the output cell voltage rises, the efficiency increases proportionately. This phenomenon is specifically useful in transportation applications. For instance, the load that needs to be provided to a fuel cell vehicle varies continuously throughout its drive cycle and a significant amount of the drive time may be spent under partial or low load conditions. Hence, in such cases, fuel cells provide an added advantage of higher operation efficiencies unlike the conventional internal combustion engines. Higher efficiency would also lead to other benefits including efficient fuel consumption as well as overall vehicle performance and lifetime.

A central advantage of fuel cell technologies that have provided them with the increased attention across the globe in the recent past comprises their environmentally benign nature of power generation. Since no harmful emissions arise as a result of power generation, fuel cells provide an environmentally friendly as well as efficient method of energy production. For example, hydrogen fuel cells results in only water as the output emission and ammonia fuel cells generally result in water and nitrogen as the reaction products. Moreover, as compared to conventional heat engine power generation technologies, fuel cells result in lower thermal emissions. Heat engine power plants entail high operating temperatures that result in both higher irreversibilities and rejection of waste heat to the atmosphere. Although fuel cell operation is also associated with heat generation due to the exothermic reactions, the amount of heat generated and the associated temperatures are generally much lower than heat engines. However, some type of fuel cell technologies that operate at high temperatures such as SOFCs, require operating temperatures of nearly 800°C for efficient operation. Nevertheless, the amount of waste heat emitted into the atmosphere is much lower. In several heat engine-based power plant facilities where lake or marine water is utilized for cooling, there have been reports of damage to marine life owing to excessive heat dumped into these water bodies. As compared to conventional thermal power plants, fuel cell power generation facilities result in significantly lower thermal emissions.

Another advantage of fuel cell technologies comprises favorable dynamic load response capability. Under dynamic loadings, fuel cell technologies entail swift adaptation characteristics. In case of a variation in output load, a reliable and fast response can be achieved. As fuel cell power generation is only limited by ionic or mass transport phenomenon, the response time associated with adapting to a given change in load is only hindered by the flow of reactants, which can also be designed to provide swift inputs and outputs. The response time of only a few milliseconds has been found to be associated with low-temperature hydrogen fuel cells, when adapting to a different load requirement. Also, large fuel cell power generation facilities with power output capacities of several MW have also been reported to entail nearly 0.3–1 s response time to a given change in load [10]. In addition to this, fuel cell-based power plants were also found to meet the standard requirements for load response that is necessary for bus applications as well as passenger vehicles along with stationary power generation facilities.

The absence of excessive noise as well as vibrations is another advantage of fuel cell technologies. Unlike heat engines that include several rotating and moving parts and equipment, fuel cells primarily operate without the need for rotating equipment with the exception of some auxiliary equipment that are utilized for reactant or oxidant supply. Furthermore, owing to their efficient, environmentally benign and lower requirements of disposal of thermal energy, fuel cells can be satisfactorily deployed in stationary on-site energy production facilities. Hence, as on-site power generation is made possible with such systems, reliance on transmission grids and distribution systems can be reduced. In the current energy generation infrastructure, massive thermal power plants are built near suitable locations where cooling towers can be easily deployed and the power generated is transmitted via transmission lines from the point of generation to the point of usage after going through a series of voltage drops and rises through transformers. Each step in such energy generation infrastructure includes irreversibilities. For instance, once the chemical energy of the fuel (oil, natural gas, etc.) is converted into electrical energy with an efficiency of approximately 20%–30%, the generated electrical energy is then passed through a transformer to make the voltage suitable for transmission. Once the suitable voltage is achieved, the transmission stage is proceeded which is associated with its own energy and thermal losses along the transmission lines that can extend up to several kilometers in length. After transmission to the point of usage, the voltage is again dropped to suitable levels that can be utilized in normal households. This is also associated with its respective energy losses as well as irreversibilities. Hence, fuel cell systems provide an opportunity to generate on-site power without any excessive noise, emissions, or cooling systems. Moreover, the absence of noise during power generation in fuel cells is also beneficial in transportation applications where the noise pollution associated with urban transportation can be reduced.

In addition to this, fuel cells also entail the potential to develop cogeneration systems. Owing to the exothermic nature of electrochemical reactions, the generated heat can be utilized for useful purposes. Although the waste thermal energy obtained from fuel cell systems is not at high temperatures as in the case of heat engines, the generated heat can be satisfactorily employed for producing heating or hot water. Generally, the fuel cell stacks have embedded coolant flow channels that continuously pass over each cell to absorb the generated heat and dispose it outside the stack. Also, in low-temperature fuel cells such as PEM fuel cells, the membrane activity significantly deteriorates if temperatures of over 80°C are achieved. Hence, the continuous coolant (generally water) supply ensures that all cells are below this threshold at all times. With the possibility of implementing cogeneration fuel cell systems, their efficient operation becomes more promising. Cogeneration aids in achieving higher overall efficiencies; hence, already efficient fuel cell systems can be operated with higher overall efficiencies. Also, fuel cells are known to operate with high efficiencies regardless of the operation load as well as power outputs. Both systems with high power outputs of several MWs to low power outputs of several watts have been found to operate with satisfactorily high efficiencies. Moreover, higher reliability has been associated with fuel cell power generation systems as compared to conventional heat engine power plants. The conventional heat engine power plants include several moving and rotating parts that require regular maintenance and inspection. However, as fuel cell systems entail minimum moving parts, limited maintenance, inspection, and supervision is required. An ideal example that demonstrates this high reliability characteristic is the first commercial fuel cell system that surpassed the maintenance reliability requirement for stationary power plants [7]. Also, fuel cell failure in space applications was found to occur after 80,000 h of operation, which was associated with a water blockage issue due to contaminants. Hence, the primary reasons of fuel cell failures generally comprise faults in the auxiliary equipment such as sensors, valves, pumps, cooling systems, etc.

2.8 Fuel cell classification

Since the invention of fuel cells, several new types of electrochemical energy conversion devices have been developed. Various methodologies have been employed for extracting electrical energy via electrochemical reactions through different types of fuels. Although hydrogen is currently the prominent fuel, other types of fuels including hydrocarbons have also been investigated for high-temperature fuel cell applications. In addition, different types of electrolytes have been introduced that can be utilized in fuel cells for ion transfer during the electrochemical reactions. Furthermore, temperature has also become a major factor that determines the operating phenomenon of fuel cells. Both high- and low-temperature fuel cells have been investigated extensively in the recent past. Fuel cells can thus be classified according to different criteria discussed in the following sections.

2.8.1 Electrolyte-based classification

As discussed in the previous sections, electrolytes are vital components of fuel cell technologies. For ion transfer, different types of electrolytes can be employed according to the fuel cell operating principle. Fig. 2.7 shows the utilization of different types of electrolytes in different fuel cell types. First, the PEM fuel cell shown on the far right entails the usage of a membrane electrolyte, which allows the passage of positively charged H^+ ions (also known as protons). Hence, in PEM fuel cells, the electrochemical reactions are accompanied by continuous flow of ions through an acidic membrane electrolyte. Second, alkaline fuel cell (AFC) technology entails the passage of negatively charged hydroxyl (OH^-) ions through an electrolyte. Currently, such electrolytes are composed of anion exchange membranes. The cathodic reaction of oxygen and water molecules generate hydroxyl ions, which are transferred through the electrolyte to the anodic side where the half-cell anodic electrochemical reaction of the fuel occurs. Third, the SOFCs

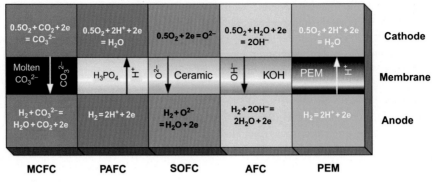

FIG. 2.7

Schematic representation of the classification of fuel cells.

entail the usage of a ceramic solid oxide electrolytes that allow the passage of negatively charged O^{2-} or positively charged H^+ ions depending on the functionality. The electrolytes allowing passage of O^{2-} ions are termed as oxygen anion-conducting electrolytes and the electrolytes allowing H^+ ion transfer are known as proton-conducting electrolytes.

The phosphoric acid fuel cell (PAFC) includes the usage of acid electrolytes, which also allows the passage of positively charged H^+ ions. Generally, phosphoric acid has been found to be a suitable acid electrolyte for fuel cell applications. However, investigation for other more suitable acid electrolytes is underway. The molten carbonate fuel cell (MCFC) entails the usage of a molten salt electrolyte that contains carbonate salts, hence, allowing the passage of negatively charged CO_3^{2-} ions from the cathodic to anodic side. At the cathode, the oxygen molecules are allowed to react with carbon dioxide molecules to form carbonate ions, which are transferred to the anode and sequential electrochemical reactions proceed.

2.8.2 Temperature-based classification

Different types of fuel cells described above entail different operating temperatures according to the limitations of the electrolyte, electrode materials and functionalities. The SOFC electrochemical energy conversion methods utilize high temperatures of around 1000°C and are thus classified as high-temperature fuel cells. Also, MCFC include the usage of molten salts and hence require high temperatures of nearly 650°C. These type of fuel cells are also classified as high-temperature fuel cells. The PAFC can operate at varying conditions and is generally referred to as a medium-temperature fuel cell considering that it operates at temperatures of nearly 200°C. Further, the PEM and AFC operate at low temperatures of 60–80°C and are thus classified as low-temperature fuel cell technologies.

2.8.3 Reactant-based classification

Apart from classification based on operating principles and conditions, fuel cells are also classified according to the type of reactants utilized. Hydrogen (H_2) fuel cells denote the utilization of hydrogen as fuel. Ammonia (NH_3) fuel cells represent the family of different types of fuel cells operating with ammonia fuel. Further, H_2-O_2 cells denote the usage of hydrogen fuel with pure oxygen as the input at the cathode side. However, H_2-air cells are fuel cells that operate with hydrogen fuel and air as the cathode input. As air contains nearly 21% oxygen, the O_2 molecules required for the cathodic reaction are obtained from air. Generally, the remaining nitrogen gas exits the cell as input. Moreover, ammonia-air (NH_3-air) fuel cells entail the input of ammonia fuel at the anode along with air input at the cathode. In addition to this, methanol has also been investigated as fuel for fuel cells in the recent past and such type of fuel cells are classified as methanol fuel cells. Other types of fuel cells that can be mentioned are the hydrogen-bromine and hydrogen-chlorine fuel cells. Also, recent studies have exerted efforts toward investigating the usage of new types of fuels including ammonia borane, hydrazine, and ammonium carbonate.

2.8.4 Ion transfer-based classification

Half-cell electrochemical reactions are integral parts of any fuel cell technology. The type of half-cell reactions that occur at the anode and cathode depend on the positively and negatively charged ions that form at the electrodes. Hence, fuel cells can also be classified according to the type of ions that transfer from one electrode to the other during the electrochemical reactions. The PEM fuel cells, which entail the transfer of positively charged H^+ ions, are also classified as cation transfer fuel cells. Such cells are associated with slow ORR at the cathode. These half-cell reduction reactions require the usage of expensive electro-catalysts that can aid in the enhancement of electrochemical reaction kinetics as well as improve the electrochemical activity. The usage of such electro-catalysts has been the primary reason for high costs associated with fuel cell technologies. Water is generally formed as a result of the electrochemical reactions in fuel cells. If the formed water molecules are not continuously removed from the cell, water flooding can occur in the fuel cell. Especially at the cathode, where oxygen reacts electrochemically to form water, flooding can be a major concern that deteriorates the fuel cell performance considerably. This can also be attributed to the lower diffusion coefficient of oxygen, which hinders the oxygen molecules from reaching the active sites satisfactorily in the presence of excess water. The abovementioned challenges are primary hindrances in the flourishment of PEM fuel cell technologies.

In addition, the AFC that include the transfer of negatively charged OH^- ions are classified as anion transfer fuel cells. Further, MCFCs are also classified as carbonate ion transfer cells owing to the transfer of negatively charged carbonate (CO_3^{2-}) ions through the electrolyte. Moreover, among SOFC technologies the classification is based on the transfer of either oxygen anions (O^{2-}) or hydrogen ions (H^+).

The former are often referred to as SOFC-O and the latter are generally classified as SOFC-H. The negatively charged ion transfer differs from positive ion transfer in advantageous ways. The ORR in such cells occurs at a higher reaction rate and entails netter reaction kinetics as compared to proton-conducting cells. Furthermore, the reaction product of water is formed at the anode or at the fuel supply side. Hence, in the case of hydrogen fuel cells where high diffusion coefficients are associated with fuel molecules, lower mass transport limitations occur as compared to PEM cells. In other words, both lower activation and concentration polarization losses are associated with such fuel cell types.

2.9 Closing remarks

In this chapter, the fundamentals of fuel cells are discussed along with the methods of classification. The hydrogen economy concept is introduced explaining the need, objectives, and development plans focusing on this idea of carbon-free society with hydrogen. In addition, the underlying operating phenomenon of fuel cells is described drawing similarities with other types of energy conversion devices such as batteries as well as thermal engines. The similarities of fuel cells with other electrochemical devices are discussed showing the similar energy conversion routes undertaken in both technologies. Also, the necessary components that are required for the operation of a fuel cell device are presented considering the process required to complete a given electrochemical reaction. Moreover, the chemical and physical phenomena that occur within a fuel cell during the electrochemical interactions including molecular diffusion as well as adsorption are presented. The classification of different types of fuel cells according to the type of electrolyte, electrode, operating temperature, etc., are also elucidated.

Types of fuels

In this chapter, different types of fuels that have been investigated in recent years for fuel cell applications will be described. Their comparative performances, compatibility with different types of fuel cells, as well as the advantages and disadvantages of each will be covered. A comparison of various energy-related properties of ammonia and other fuels is depicted in Table 3.1. As can be observed from the table, ammonia entails a comparatively higher hydrogen density per unit volume of $136\,kgH_2/m^3$. This is higher when compared to other fuels including hydrogen that entails $70\,kgH_2/m^3$ as well as ethanol that has a hydrogen density of $102\,kgH_2/m^3$. Ammonia also entails other favorable properties such as high volumetric density. The base energy density of ammonia is lower than other fuels; however, the reformed energy density entails the highest value. The reformed energy density incorporates the water as well as heat volume required. Furthermore, ammonia also has a higher octane number as well as compression ratio. These can be favorable depending on the type of engine that is utilized to generate power. Although current power generation engines are designed to operate with carbon-based fuels including gasoline, propane, etc., some efforts were made to develop engines that operate with ethanol fuel. Efforts are underway to develop combustion engines that can operate with pure ammonia fuel. Nevertheless, thermal engines are limited by the Carnot efficiency that is restrained by the temperature levels of the heat source and sink. Generally, the efficiencies of these engines remain in the range of 20%–30% due to various thermodynamic irreversibilities within the system. However, fuel cells are not limited by the Carnot rule as thermal energy is not directly converted to electricity; rather, chemical energy is directly converted into electrical energy. Hence, increased attention toward the utilization of fuel cells is being paid across the globe with various different types of fuels.

3.1 Hydrogen fuel

Hydrogen is the most prominent fuel for nearly all fuel cell technologies. Fuel cells have gained increased attention in the recent past, specifically from the transportation sector, owing to the desirable characteristics of utilizing hydrogen as fuel. It entails the lowest molecular weight and one of the highest lower heating values per unit mass. Furthermore, hydrogen is also sometimes referred to as one of the most abundant substances on earth considering the various different chemical species and

Ammonia Fuel Cells. https://doi.org/10.1016/B978-0-12-822825-8.00003-7

Table 3.1 Comparison of ammonia with other fuels.

Fuel	H$_2$ density (kg H$_2$/m^3)	Volumetric density (kg/L)	Higher heating value (MJ/kg)	Energy density (MJ/L)	Reformed energy density (MJ/L)	Octane number	Maximum compression ratio
Ammonia	136	0.76	22.5	15.3	13.6	130	50:1
Hydrogen	70	0.07	141.9	9.9	9.9	130	–
Ethanol	102	0.78	29.7	23.4	9.1	109	19:1
Methanol	98	0.79	22.7	17.9	10.2	109	19:1
Propane	106	0.51	48.9	29.4	8.6	120	17:1
Gasoline	110	0.70	46.7	36.2	9.2	86–94	10:1

Source: From Refs. [79,80].

compounds it forms with several other elements. It is an odorless gas and entails a nontoxic nature. However, it is highly combustible in the presence of oxygen molecules. This is one of the major safety issues that is currently associated with the usage of hydrogen. Specifically, in the transportation sector, fuel cell vehicles are considered to entail high risks for passengers due to their highly flammable nature. Moreover, hydrogen has a boiling point of $-252.87°C$ at ambient conditions. Hence, compression and storage of hydrogen can be challenging.

3.1.1 Hydrogen for fuel cells

In hydrogen fuel cell systems such as fuel cell vehicles or stationary power generators, either very low-temperature cryogenic techniques would be needed to liquefy the produced hydrogen or high pressures would be needed to store sufficient amounts of hydrogen gas. Fig. 3.1 depicts the effect of pressure on the density of hydrogen at different temperatures. As can be observed from the figure, the density of hydrogen at an ambient pressure is considerably low ($0.08209 kg/m^3$). This value is lower than the density of air at standard conditions ($1.2 kg/m^3$). Further, as the pressure of hydrogen gas is increased, although a considerable rise is observed, the values of density are comparatively low.

For instance, at a high pressure of 100 bar, the density of hydrogen is nearly $7.8 kg/m^3$ at an ambient temperature of 25°C. This value is observed to increase to merely $9.2 kg/m^3$ at a low temperature of $-25°C$. Moreover, if high pressures of 800 bar are utilized, the density of hydrogen gas increases to $42.9 kg/m^3$ at 25°C and to $48.36 kg/m^3$ at a temperature of $-25°C$. Hence, if gaseous hydrogen fuel is compressed to high pressures of 800 bar, 0.0429 kg of fuel could be stored in a volume of 1 L. However, to provide a comparison of these density values, Fig. 3.2 depicts the variation of the density of air with a change in pressure. At a pressure of 100 bar, air density reaches $117.8 kg/m^3$ at a temperature of 25°C and increases to $149.6 kg/m^3$ at a low temperature of $-25°C$. Further, if a high pressure of 800 bar is utilized, air density increases to $560.8 kg/m^3$ at an ambient temperature

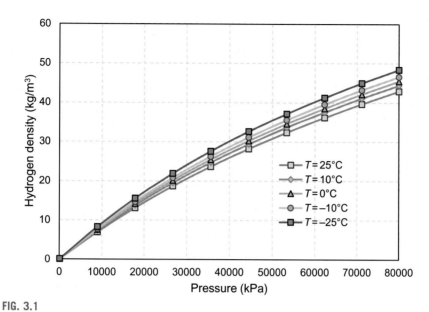

FIG. 3.1

Effect of pressure on density of hydrogen at various temperatures.

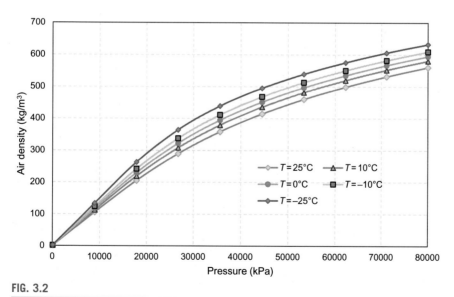

FIG. 3.2

Effect of pressure on density of air at various temperatures.

of 25°C. Thus, more than 10 times the amount of hydrogen could be stored at the same pressures if air is considered as a comparative example. However, this value increases exponentially when compared to fuels such as ammonia that can liquefy at comparatively low pressures as well as temperatures.

Nevertheless, although hydrogen entails a low volumetric density, it provides the most suitable fuel cell performances. As compared to other fuel cell fuels, hydrogen provides the highest open-circuit potentials as well as peak power densities. Moreover, different types of hydrogen fuel cells provide different potentials as well as power densities. For instance, the comparison of open-circuit potentials between different hydrogen fuel cells is depicted in Fig. 3.3. The PEM fuel cell, which is also the most prominent type of fuel cell currently, provides an open-circuit potential of nearly 1.1 V. Theoretically, the Nernst potential can be evaluated to be nearly 1.2 V; however, owing to irreversibilities, the theoretical value is not achieved. Following the PEM fuel cell, the PAFC type cell is also observed to provide high open-circuit potentials. This is followed by the SOFC and AFC that provide nearly 1.0 V. However, when a comparison of power density is made as depicted in Fig. 3.4, PEM hydrogen fuel cells provide a considerably higher density of $6.5 \, kW/m^3$. Furthermore, this is followed by the MCFC technology, which entails a power density of $2.6 \, kW/m^3$. Also, PAFC and SOFC-type cells generally entail power densities of 1.9 and $1.5 \, kW/m^3$, respectively [10,11]. As can be depicted from Fig. 3.4, hydrogen provides significantly better fuel cell performance when utilized with the PEM technology.

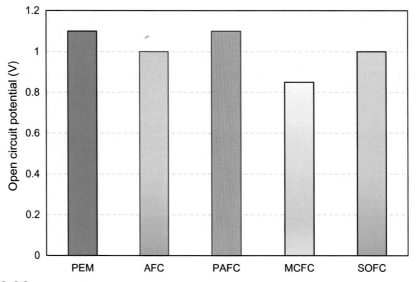

FIG. 3.3

Comparison of open-circuit potentials of different hydrogen fuel cells.

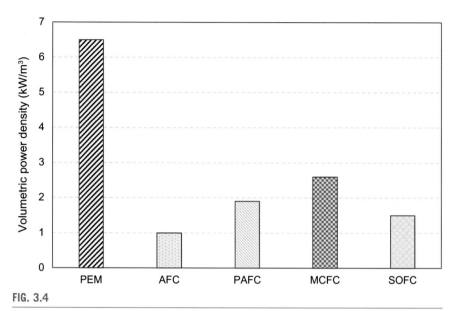

FIG. 3.4

Comparison of power densities of different hydrogen fuel cells.

This can be attributed to several reasons. First, in recent years, several efforts toward the development of commercially viable PEM fuel cell technologies have been exerted. This has led to the introduction of various innovative, cost-effective, and high-performance PEM fuel cell systems. Such systems have been made more compact and efficient over years of development by various institutions, organizations, and companies across the globe. However, other types of fuel cells including AFC and SOFC have started to gain increased interests and their performances in terms of output voltages as well as power densities have been observed to improve over the years.

Depending on the type of fuel cell, the electrochemical reactions of hydrogen vary. In PEM fuel cells, the electrochemical oxidation of hydrogen that occurs at the anode can be denoted as

$$H_2 \rightarrow 2H^+ + 2e \tag{3.1}$$

Also, the cathodic reaction that occurs in PEM fuel cells between positively charged hydrogen ions (also known as protons) is represented as

$$\frac{1}{2}O_2 + 2H^+ + 2e \rightarrow H_2O \tag{3.2}$$

Further, the overall reaction for the fuel cell can be written as

$$H_2 + \frac{1}{2}O_2 \rightarrow H_2O \tag{3.3}$$

However, in the case of AFC the anodic reaction of hydrogen molecules occurs with negatively charged hydroxyl (OH^-) ions. This can be represented as

$$H_2 + 2OH^- \rightarrow 2H_2O + 2e \tag{3.4}$$

Moreover, the cathodic reaction in hydrogen fueled AFC also varies. The oxygen molecules that input the cathode chamber react with water molecules to form the hydroxyl ions necessary to continue to anodic reactions. This cathodic interaction can be denoted as

$$\frac{1}{2}O_2 + H_2O + 2e \rightarrow 2OH^- \tag{3.5}$$

In addition to this, in SOFC the anodic interaction of hydrogen molecules comprises a half-cell reaction with negatively charged O^{2-} ions. The anodic reaction of SOFC can be represented as

$$H_2 + O^{2-} \rightarrow H_2O + 2e \tag{3.6}$$

Further, the negative oxygen ions are formed at the SOFC cathode according to the following reaction:

$$\frac{1}{2}O_2 + 2e \rightarrow O^{2-} \tag{3.7}$$

Furthermore, the PAFC technology included electrochemical reactions similar to PEM fuel cells owing to the same nature of acidic electrolyte and transfer of positively charged hydrogen ions or protons. MCFC involves the interaction with carbonate ions. The anodic reaction for MCFC fuel cell type can be denoted as

$$H_2 + CO_3^{2-} \rightarrow H_2O + CO_2 + 2e \tag{3.8}$$

The corresponding cathodic reaction for MCFC can also be represented as

$$\frac{1}{2}O_2 + CO_2 + 2e \rightarrow CO_3^{2-} \tag{3.9}$$

The overall reaction, however, remains the same in different fuel cell types where hydrogen and oxygen interact to form water molecules as denoted by Eq. (3.3).

3.1.2 Hydrogen production methods

There are several methods of hydrogen production. Some methods such as steam reforming are commercially established and some methods such as water electrolysis have been recently commercialized at small scales. Steam methane reforming (SMR) is the primarily utilized method currently and entails various environmental detriments. This process involves the reaction of methane with steam at temperatures of nearly 700–1100°C in the presence of nickel catalyst, which can be written as

$$CH_4 + H_2O \rightleftharpoons CO + 3H_2 \tag{3.10}$$

Further, carbon monoxide is generally converted to carbon dioxide and more hydrogen gas by a water gas shift reaction:

$$CO + H_2O \rightleftharpoons CO_2 + H_2 \tag{3.11}$$

As can be observed from the above reactions, the SMR process entails large amounts of CO_2 emissions and is thus environmentally detrimental. Thus, other cleaner methods of hydrogen production are being investigated that are more environmentally benign.

Water electrolysis is another hydrogen production method that can provide solutions to the environmental detriments associated with the current hydrogen production based on methane reforming. Electrolysis entails the consumption of electricity to dissociate water molecules to produce hydrogen and oxygen. This can be depicted by the overall reaction for any given water electrolyzer:

$$H_2O + P_{in} \rightarrow H_2 + \frac{1}{2}O_2 \tag{3.12}$$

where P_{in} denotes the electrical power input. Since this method of hydrogen production relies primarily on electrical energy, renewable and clean energy resources such as solar, wind, and hydro can be used to generate electrical power, which can then be used to produce hydrogen electrochemically. As can be observed from Eq. (3.12) water electrolysis operates reversible to a fuel cell. In a fuel cell, where hydrogen and oxygen react to produce water, an electrolyzer entails the dissociation of water molecules to produce hydrogen as well as oxygen. A fuel cell provides an output of electrical power whereas an electrolyzer requires an electrical power input. Hence, similar to fuel cells, different types of electrolyzers exist that are classified according to the electrolytes, operating temperatures, and ion transfer mechanisms. Most types are common to both such as PEM, AFC, and solid oxide cells. Moreover, the above described water electrolysis process is also being investigated to develop into a photoelectrochemical process. In this method, photoactive materials are coated onto the electrolyzer electrodes and both electrical energy and photonic energy are input to the reactor. When solar radiation strikes the photoactive electrodes, the process of dissociation of water molecules is accelerated due to the photosensitive coatings.

Another method of hydrogen production includes thermochemical methods of reforming as well as gasification. Specifically, the usage of biomass gasification (BG) for hydrogen production has attained increased interest in recent years. The general biomass dissociation can be denoted as

$$a C_b H_c O_d + e H_2O + Heat \rightarrow f H_2 + k Tar + i CH_4 + g CO_2 + h CO + j C \tag{3.13}$$

In the above reaction, the biomass is represented by the first term on the left-hand side. The coefficients of carbon, hydrogen, and oxygen atoms can vary according to the type of biomass. Each type of biomass entails different compositions.

Another type of hydrogen production comprises thermochemical cycles (TCs) such as the CuCl cycle. In this type of hybrid cycle, both electrical and thermal energy input are provided. However, the electrical energy input required is lower than the general water electrolysis. The chemical reactions associated with a four-step cycle entailing the CuCl cycle are

$$2CuCl_2(s) + H_2O(g) \rightarrow Cu_2OCl_2(s) + 2HCl(g) \tag{3.14}$$

$$Cu_2OCl_2(s) \rightarrow 0.5\,O_2(g) + 2CuCl(l) \tag{3.15}$$

$$2CuCl(aq) + 2HCl(aq) \rightarrow H_2(g) + 2CuCl_2(aq) \tag{3.16}$$

$$CuCl_2(aq) \rightarrow CuCl_2(s) \tag{3.17}$$

where the cyclic reactions take material inputs of water and provide outputs of hydrogen (H_2) as well as oxygen (O^2).

Furthermore, plasma arc decomposition (PAD) is another method of hydrogen production that entails the decomposition of hydrogen containing compounds to form hydrogen via generation of plasma. For instance, methane can be decomposed according to the equation

$$CH_4 \rightarrow C + 2H_2 \tag{3.18}$$

In this method, dielectric barrier discharge is utilized to generate plasma that results in the ionization of several species as well in the decomposition of various compounds owing to the attainment of high temperatures.

Moreover, a simple yet energy intensive process of hydrogen production from water includes water thermolysis. The general chemical reaction for this process can be denoted as

$$H_2O + Heat \rightarrow H_2 + \frac{1}{2}O_2 \tag{3.19}$$

The above process is highly energy intensive as temperatures of nearly 2500 K would be necessary to achieve sufficient dissociation amounts. For example, nearly 64% of the water would be expected to dissociate into hydrogen and oxygen at a temperature of 3000 K and 1 atm, Hence, to achieve higher conversions, harsher environments would be needed.

In addition, other biochemical methods have also been investigated for the production of hydrogen. Biophotolysis comprises one such method. In this technique, the natural photosynthetic reaction is utilized in the presence of enzymes that result in the following chemical reactions

$$6H_2O + 6CO_2 + hv \rightarrow C_6H_{12}O_6 + 6O_2 \tag{3.20}$$

$$C_6H_{12}O_6 + 6H_2O + hv \rightarrow 6CO_2 + 12H_2 \tag{3.21}$$

A comparison of the energy as well as exergy efficiencies of the above discussed hydrogen production routes is depicted in Fig. 3.5. As can be observed, the highest energy efficiencies are entailed with the SMR hydrogen production route comparatively. This explains the widespread usage of the technology globally to produce

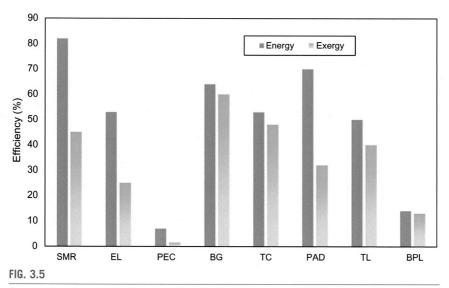

FIG. 3.5

Comparison of energy and exergy efficiencies of hydrogen production through BG, biomass gasification; BPL, biophotolysis; EL, water electrolysis; PAD, plasma arc decomposition; PEC, photoelectrochemical water electrolysis; SMR, steam methane reforming; TC, thermochemical cycles; TL, thermolysis.

hydrogen. The energy efficiency of this route is around 82%. However, in terms of the exergy efficiency, BG is the hydrogen production route with a comparatively higher value of approximately 60%. The energy efficiency of this route is nearly 64%. Moreover, the water electrolysis (EL) route entails an energy efficiency of around 53%. However, the associated exergy efficiency lies in the comparatively lower range (25%). The photoelectrochemical (PEC) route of hydrogen production entails both low energy as well as exergy efficiencies of 7% and 1.5%, respectively. Furthermore, the hybrid TC hydrogen production route entails average overall efficiencies of nearly 53% and 48%. Also, the PAD route is associated with a high energetic efficiency of around 70% but entails a comparatively lower exergy efficiency of 32%. In addition to this, the thermolysis (TL) route has an energy efficiency of about 50% and an exergy efficiency of nearly 40%. The biophotolysis (BPL) hydrogen production route entails comparatively lower efficiencies of 14% energetically and 13% exergetically.

In addition to the efficiencies, several other factors need to be considered while assessing any fuel production methodologies. The primary factor generally includes the cost per unit quantity of fuel produced. Hence, among the methods described above, if the most cost-effective technique for hydrogen production is utilized, the techno-economic feasibility of hydrogen will be greatly improved. A comparison of the unit cost of production for several hydrogen production routes is depicted in Fig. 3.6. As can be observed from the figure, the PEC route is comparatively more

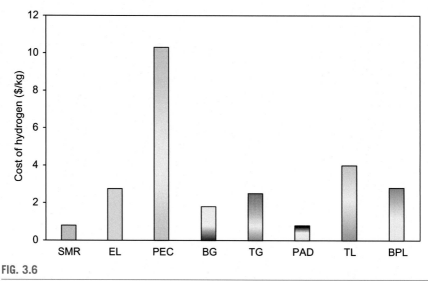

FIG. 3.6

Comparison of production cost of hydrogen production through BG, biomass gasification; BPL, biophotolysis; EL, water electrolysis; PAD, plasma arc decomposition; PEC, photoelectrochemical water electrolysis; SMR, steam methane reforming; TC, thermochemical cycles; TL, thermolysis.

cost intensive with a production cost of nearly 10.3 $/kg. This can be attributed to the usage of photosensitive materials such as titanium oxides or iron oxides for coating the electrodes. The electrolyzer electrodes need to be coated with such materials to generate photonic energy that is required for a photoelectrochemical process. After this route, the thermolysis route entails a comparatively higher cost of 4 $/kg. This is followed by BPL (2.8 $/kg), EL (2.75 $/kg), and TC (2.5 $/kg). The lowest production of nearly 0.8 $/kg is associated with the SMR process. This explains the usage of this method to meet the majority of the world's hydrogen demand. Although the method is associated with various environmental detriments owing to large amounts of carbon emissions, it entails a low-cost technology that is also associated with a high energy efficiency. Thus, it is essential to develop renewable and clean methods of hydrogen production that are both competitively efficient as well as cost-effective.

3.1.3 Hydrogen storage methods

One of the key challenges that hinders the flourishment of hydrogen fuel is associated with its low volumetric density. Although hydrogen contains a higher energy content per unit mass as compared to other fuels, it entails considerably lower energy content per unit volume. Owing to this reason, several methods to store hydrogen have been introduced in the recent past and many other types of methodologies are currently under investigation. Fig. 3.7 depicts a summary of the different types of hydrogen

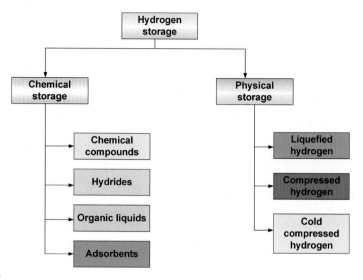

FIG. 3.7

Schematic representation showing different methods of hydrogen storage.

storage methods. The general classification can be made on the type of storage means that can be classified as physical or chemical storage. Physical storage entails the storing of hydrogen itself without any external chemical interactions with other species or compounds. Physical storage can be further classified into liquefaction, compression, or cryo-compression. Liquefaction includes converting the gaseous hydrogen into liquid hydrogen through a series of multiple physical processes. Liquefaction can be useful in the hydrogen storage industry due to the higher density associated with the liquid hydrogen as compared to gaseous hydrogen. Thus, more amount of mass can be stored per unit volume. Next, the compression type of physical storage involves compressing the hydrogen gas to high pressures without liquefying. Currently, high pressures of nearly 700 bar are also utilized in the compression hydrogen storage technique. This is a common technique and is comparatively less complicated than the liquefaction method. Furthermore, the cryo-compression technique also involves the cold storage of hydrogen. However, this technique entails various advantages as compared to the liquefaction technique. In this method of hydrogen storage, when the liquid hydrogen vaporizes due to thermal energy absorption from the atmosphere, the pressure inside the storage tank is allowed to reach high values to nearly 350 bar, thus allowing a higher driving range. This method was tested with hybrid electric vehicles and it was found to provide a driving range of nearly 1050 km.

The chemical means of hydrogen storage involve the utilization of hydrogen containing chemical species or chemical interaction of hydrogen with other chemical species that allow to increase the storage density. This method can also be divided into various subcategories. Hydrogen containing chemical compounds is the first type of chemical storage method. These compounds can include hydrocarbons, alcohols, ammonia, etc. In case of hydrocarbons, the formed chemical compounds would have to be reformed to extract usable hydrogen. Also, the high-temperature fuel cells such as solid oxide fuel cells (SOFCs) also allow direct internal reforming of hydrocarbon fuels. For instance, methane has been investigated to provide satisfactory fuel cell performance when used as fuel in SOFCs. Furthermore, alcohols such as ethanol and methanol have also been investigated. These can also be utilized directly as fuels in fuel cell applications. However, their performances are lower as compared to direct hydrogen fuel cells. In addition, ammonia has been investigated in the recent past as a direct fuel for fuel cells or as an indirect source of hydrogen for fuel cells. The proceeding chapters will discuss in detail various aspects of different types of ammonia fuel cells. Next, the chemical means of hydrogen storage include hydrides. These can also be divided into either metallic or nonmetallic hydrides. The metallic hydrides include magnesium hydride, lithium hydride, sodium-aluminum hydride, etc. These chemical compounds have been investigated owing to the hydrogen atoms they entail that can be used to obtain hydrogen gas whenever required. They can sometimes be liquids that can be readily fueled. However, metallic hydrides entail strong hydrogen bond energies and thus require high temperatures and thermal energy inputs to extract hydrogen. The required input energy may be reduced through the utilization of alloys such as $NaBH_4$ or $LiBH_4$. Such compounds entail lower bond energies and comparatively weaker bonds that make it easier to extract hydrogen. Several studies have been conducted in the recent past on investigating hydrogen storage in the form of various types of metallic alloy hydrides. There also exist nonmetallic hydrides such as hydrazine and silicon hydrides. Specifically, hydrazine has been found to entail promising properties and has been suggested by some to be utilized as a replacement for hydrogen. Next, organic liquids have also been tested as a means of chemical hydrogen storage. Hydrogenation of organic liquids is performed to store hydrogen chemically and dehydrogenation is conducted when hydrogen extraction is desired. Cycloalkanes were initially investigated extensively as promising organic liquids for hydrogen storage owing to their comparatively high capacity of hydrogen of nearly 6%–8% by mass. However, other chemical compounds including N-ethylcarbazole and dibenzyltoluene have been identified as appropriate organic liquids that can be employed as hydrogen carriers. The dibenzyltoluene entails a wide range of temperatures between the freezing (−39°C) and boiling points (390°C). Also, it is associated with a relatively high storage percentage of hydrogen of 6.2% by weight. Another material includes formic acid, which entails a storage capacity of nearly 4%. Moreover, the organic liquid route of hydrogen storage has been found to entail higher overall hydrogen extraction efficiencies as compared to obtaining hydrogen via chemical compounds such as hydrocarbons. Next, the method of physisorption of hydrogen on different materials is also being investigated where hydrogen can be adsorbed on certain materials and desorption can be conducted through varying the temperatures or pressures whenever hydrogen extraction is required. Different types of materials have been tested for hydrogen adsorption that includes carbon-based materials such as carbon nanotubes as well as zeolites, alumina, and silica compounds.

3.2 Alcohol fuels

The alcohol family of chemical compounds such as ethanol and methanol has been utilized as direct fuels in fuel cell applications. This chapter presents such fuels and discusses their performances and latest developments.

3.2.1 Ethanol fuel

Ethyl alcohol or ethanol is a class of alcohol compounds that has been used extensively in various applications. Their usage in alcoholic beverages is well known. However, ethanol has also been considered as a fuel additive as well as a stand-alone fuel for vehicles. In 1978, a car was developed by Fiat that operated via ethanol fuel instead of the conventional gasoline or diesel fuel used by vehicles. However, this technology relied on fuel combustion rather than through electrochemical fuel cells. Direct ethanol fuel cells (DEFC) entail the utilization of ethanol fuel directly in place of hydrogen in fuel cells. Ethanol has been reported to entail a relatively high energy density of 8 kWh/kg. Generally, direct ethanol fuel cells utilized the conventional PEM membrane electrolytes that are coated with platinum black catalysts. However, other challenges including CO_2 formation as well as slow electrochemical oxidation reaction rates of ethanol needed to be solved. Alkaline electrolyte entailing fuel cells such as anion exchange membrane-based cells, however, provided better performances. Specifically, methanol electrooxidation was found to entail more favorable polarization behaviors in alkaline electrolyte-based cells than in proton exchange membrane-based cells. Moreover, alkaline electrolytes also provide the opportunity to utilize other nonnoble metal catalysts that have lower costs unlike the conventional platinum black catalyst [21]. Furthermore, hybrid alkaline and acidic medium-based ethanol fuel cells have also been developed, which generally utilize cation exchange membrane electrolytes and mixtures of ethanol and alkaline solutions. Ethanol has been identified to be a better fuel as compared to methanol owing to lower toxicity and a well-established production infrastructure [13].

The overall electrochemical reaction of DEFCs can be denoted as

$$CH_3CH_2OH + 3O_2 \rightarrow 2CO_2 + 3H_2O \tag{3.22}$$

The cathodic and anodic reactions in case of proton exchange membrane-based DEFC can be written, respectively, as

$$CH_3CH_2OH + 3H_2O \rightarrow 2CO_2 + 12H^+ + 12e \tag{3.23}$$

$$3O_2 + 12H^+ + 12e \rightarrow 6H_2O \tag{3.24}$$

As can be observed from Eq. (3.2), the formation of CO_2 at the anode associates a disadvantage with DEFC owing to carbon emissions during fuel cell operation. The overall electrochemical reaction for a hybrid alkaline and acidic DEFC can be written as

$$CH_3CH_2OH + 12NaOH + 6H_2O_2 + 6H_2SO_4 \rightarrow 2CO_2 + 21H_2O + 6Na_2SO_4 \qquad (3.25)$$

As shown in Eq. (3.25), such types of DEFC entail the utilization of alkaline solutions of sodium hydroxide (NaOH), sulfuric acid (H_2SO_4) as well as H_2O_2. The cathodic reaction can be expressed for this type of fuel cell as

$$6H_2O_2 + 12H^+ + 12e \rightarrow 12H_2O \qquad (3.26)$$

At the anode, first the NaOH entails the Na^+ and OH^- ions:

$$12\,NaOH \rightarrow 12Na^+ + 12OH^- \qquad (3.27)$$

Also, the interaction of ethanol with the OH^- ions can be expressed as

$$CH_3CH_2OH + 12OH^- \rightarrow 2CO_2 + 9H_2O + 12e \qquad (3.28)$$

Moreover, Na^+ ions can transfer via the membrane electrolyte that allows the passage of cations to reach the cathodic side. A list of several studies performed on the investigation of DEFC is given in Table 3.2. The fuels utilized and the corresponding power densities obtained at the operating temperatures used are provided from different studies reported in the literature. In addition, as the type of cathodes and anodes used plays a significant role in determining the performance of the fuel cell, these have also been listed alongside each study.

As can be observed from Table 3.2, the DEFC with input fuel of ethanol and sodium hydroxide with a cathode material of platinum on carbon, anode material of palladium-nickel on carbon and electrolytes comprising Nafion membrane have been reported to provide comparatively higher peak power densities of nearly 240 and 360 mW/cm^2. Furthermore, a comparatively lower performance has been reported to be associated with an ethanol and potassium hydroxide input fuel entailing an Fe-Co cathode and Pd$_2$Ni$_3$/C anode with an anion exchange membrane electrolyte. However, there are several other factors that can affect the performances of DEFC in terms of their peak power densities. These include the method of fabrication of catalysts, active site surface area, fuel cell design, fuel utilization, etc.

There are various challenges associated with the utilization of ethanol fuel for fuel cell applications. Fuel crossover refers to the transfer of fuel from the anodic compartment to the cathodic compartment owing to the permeability of the electrolyte. Generally, the crossover problem is associated with membrane-based fuel cells, where the membrane electrolyte allows the fuel molecules to permeate to the other side of the cell. This is disadvantageous due to the disturbances that are caused on the cathodic fuel cell side. The cathodic potentials are lowered due to undesired interactions and electrochemical reactions of the fuel at the cathode side. Also, the fuel that crossovers is considered as wasted amount of fuel as it could not

Table 3.2 Results of direct ethanol fuel cells investigated and reported in the literature.

Input	Peak power output (mW/cm^2)	Temperature (°C)	Electrolyte	Electrodes
Ethanol/NaOH	360	60	Nafion (211)	Cathode: Pt/C Anode: PdNi/C
Ethanol/NaOH	240	60	Nafion (117)	Cathode: Pt/C Anode: PdNi/C
Ethanol/KOH	60	40	AEM (A201)	Cathode: Fe-Co Anode: Ni-Fe-Co
Ethanol/KOH	44	60	AEM (A201)	Cathode: Fe-Co Anode: Pd$_2$Ni$_3$/C
Ethanol/KOH	110	80	PBI	Cathode: Co-TMPP/C Anode: RuV/C
Ethanol/KOH	58	20	AEM (A006)	Cathode: Fe-Co Anode: Pd-(Ni-Zn)/C
Ethanol/KOH	58	20	AEM	Cathode: Pt black Anode: PtRu
Ethanol/KOH	49.2	75	KOH/PBI	Cathode: Pt/C Anode: PtRu/C

Source: From Ref. [21].

be utilized at the anodic side of the cell. Further, the electro-catalysts used on the cathode side of DEFC can be deteriorated due to the products formed from the chemical interactions of ethanol. Hence, these factors considerably affect the DEFC performances as well as efficiencies. Moreover, the fuel crossover in DEFC is also affected by factors such as input fuel concentration, current densities, and temperature. These factors have been found to raise the crossover rates. When ethanol crosses over to the cathodic side, it may react with oxygen molecules to form chemical compounds such as acetaldehyde and acetic acid. These compounds can accumulate at the cathodic side and prevent oxygen molecules from reaching the catalyst surfaces, thus deteriorating the performances of DEFC considerably. In order to reduce this problem, input fuel with lower concentrations of ethanol can be used as the lower the concentration of ethanol, the lower is the crossover phenomenon. Nevertheless, the power outputs can decrease with the concentrations. Further, lower permeability membranes can also be deployed that can reduce the permeation of fuel.

Another challenge associated with ethanol fuel is its electrooxidation reaction, which comparatively entails slower kinetics. Slower kinetics of electrochemical reactions in fuel cells can lead to reduction in fuel cell performances in terms of output voltages, power outputs as well as efficiencies. One solution to enhance the reaction kinetics of ethanol electrooxidation includes utilizing higher operating temperatures. However, the electrolyte materials are not stable at very high temperatures. Hence, operating temperatures of nearly 90°C have been investigated and were found to provide comparatively higher electrooxidation kinetics. The conversion of ethanol in DEFCs has been investigated to improve when higher temperatures as well as currents were utilized. Fig. 3.8 depicts this relation between the operating temperature, cell current, and the conversion of ethanol fuel.

The operating temperatures enhance the ethanol conversion considerably. For instance, at low current densities of nearly 20 or 40 mA, the ethanol conversion is nearly doubled. In addition to this, the current density also increases the ethanol conversion significantly. For example, as the current density increases from 20 to 120 mA, the ethanol conversion rises by nearly three times at an operating temperature of 60°C. However, at an operating temperature of nearly 90°C, the ethanol conversion rises by approximately 2.3 times. Thus, increasing both the operating temperatures as well as current densities of ethanol fuel cells is desirable.

Water (H_2O) molecules are present in large numbers in ethanol fuel cells owing to the ethanol solutions utilized as well as the electrode reactions that result in the generation of water molecules as a result of the half-cell electrochemical reactions. If the presence of these molecules is not managed appropriately, they can result in significant deterioration of fuel cell performances. The flooding phenomenon can occur if excess water accumulates at an electrode. This can increase the cell

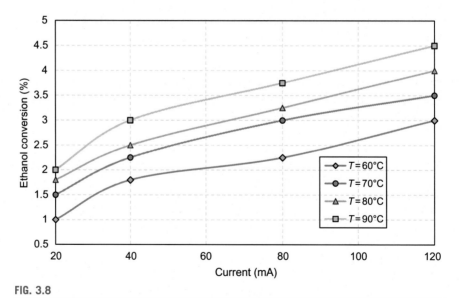

FIG. 3.8

Relationship between operating temperature, cell current, and ethanol conversion.

Data from Ref. [23].

resistances considerably and hence lower the electrochemical activity as well as overall useful output. As water is electrically conductive, short circuit of cell elements can occur if water permeation occurs from the anode to the cathode or vice versa through the membrane. Although hydration of the cell membrane is essential, excessive water absorption by the membrane is undesirable. Excessive absorption can lead to enlargement of the polymer material comprising the membrane leading to other stability issues. The excessive enlargement can also enhance the permeation of ethanol molecules leading to a further lowering of fuel cell performance. Hence, the development of efficient water management systems needs to be developed for DEFC. Moreover, the durability of fuel cells is also an important characteristic that needs to be considered. Several studies have been conducted on testing the durability of DEFC. For instance, in one study the peak power density of a DEFC was found to decrease by 15% when operated for 60 h. This test included the usage of electrodes with platinum and tin catalysts on carbon supports [24]. Further studies testing the performance deterioration with time of DEFC need to be performed.

Several combinations of different noble metals have been investigated to act as catalysts in DEFC. Palladium-based catalysts have been extensively studied as anodic catalysts for these types of fuel cells. These have shown better performances as compared to several other types of electro-catalysts and have shown enhancement in the electrochemical oxidation characteristics of ethanol fuel. This has been attributed to the greater surface area that provides more access to active sites where electrochemical reactions can occur. In addition to this, palladium catalysts have also been found to be comparatively more stable. DEFC performance with catalysts comprising palladium alloyed with other metals such as cobalt, molybdenum, and nickel has also been investigated to provide good electrochemical activities. Moreover, foam-based electrodes comprising palladium and nickel have been found to entail a few advantages as compared to conventional electrode types. These types of electrodes provide a greater amount of surface area as well as porosity owing to their three-dimensional structure. Further, the cell resistance associated with the molecular as well as ionic diffusion across the electrode is also lowered. Nevertheless, in an alkaline electrolyte comprising DEFC, nonnoble metal catalysts can also be employed, lowering the associated costs.

The other most common types of electro-catalysts utilized in DEFC comprise platinum-based catalysts. However, as these type of catalysts are primarily compatible with acidic cell types, they entail several challenges when used in DEFC. The formation of carbon monoxide deteriorates the platinum catalyst due to adsorption of molecules on the surface lowering their activities. Nevertheless, catalysts comprising several combinations of platinum and other metals have been investigated. The catalysts with comparatively higher performances were platinum-nickel-ruthenium-based catalysts on carbon support. Apart from anodic catalysts, cathodic catalysts also play an essential role. Several types of cathodic electro-catalysts have been investigated including platinum with cobalt, ruthenium, and palladium and were found to be comparatively better candidates for DEFC.

3.2.1.1 Production of ethanol fuel

Ethanol entails several applications in various industries. A significant percentage of ethanol consumption comprises making alcoholic beverages. A few decades earlier, however, ethanol was being considered as an alternative to gasoline fuel in cars and transportation vehicles. Further, the ethanol fuel-based car developed by Fiat in 1978 had opened several doors for the flourishment of ethanol fuel. Nevertheless, having a sufficient supply of ethanol fuel was one of the primary challenges that had to be addressed. If a large number of gasoline fueled vehicles were replaced with ethanol fueled vehicles, the amount of fuel required would have led to the requirement of building a massive ethanol production infrastructure. However, ethanol is currently utilized as a fuel blend or additive. Owing to the enormous increase in the usage of road transportation vehicles that are primarily comprising gasoline fueled vehicles, the consumption of ethanol as a fuel additive also increased in the recent past. Fig. 3.9 depicts the global ethanol production distribution.

As can be observed from the figure, the United States entails the highest ethanol production of nearly 58%, which constitutes approximately 15,330 million gallons of liquid ethanol per year. Followed by the United States, Brazil is the largest producer of ethanol with an annual production of nearly 7295 million gallons of liquid ethanol. This may be attributed to the high amount of usage of ethanol as a fuel additive in these countries.

In various parts of the world where ethanol is produced, the production methodologies mainly comprise three techniques. First, the synthetic method of ethanol production constitutes the reaction of steam with ethene. The required ethene is

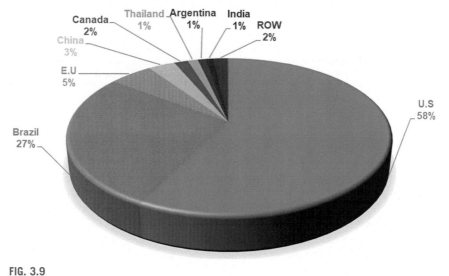

FIG. 3.9

Global ethanol production distribution.

Data from Ref. [16].

obtained from oil cracking and in the presence of suitable catalysts and reaction conditions, ethene and steam are reacted to form ethanol. The catalyst for the reaction is generally phosphoric acid based. Also, the reaction temperature of nearly 300°C as well as high reaction pressure of about 60–70 bar is generally employed for favorable reaction kinetics. However, utilizing these conditions still results in low conversion percentages of nearly 5% and continuous recycling is necessary. Moreover, in this method of ethanol production, purification is an essential step. The ethanol produced from the reaction between steam and ethene comprises nearly 4% water content, which needs to be eliminated to obtain pure ethanol. Various methods exist to separate water from ethanol. For instance, calcium oxide is employed as a drying agent to remove the low (4%) of water content. Also, benzene may be utilized to dissociate the azeotrope mixture of ethanol and water. Other new methods of purifying the final ethanol mixture have also been introduced in the recent past, one of the most promising methodologies being the usage of zeolites that absorb the water molecules and hence aid in obtaining pure ethanol with lower energy requirements as well as costs as compared to conventional purification methods. The second method of ethanol production comprises the fermentation technique. This technique has existed since several centuries and has been used commonly for producing alcohol. Even in the current era, the majority of the ethanol produced across the world relies on the technique of fermentation. In this method, organic crops or species including corn, sugarcane, maize, etc., are utilized as the feed that is fermented to produce ethanol. These are generally referred to as substrates. The type of substrate used for ethanol production in a given location depends on the availability of the type of substrate as well as economic feasibility of the area. For instance, ethanol production plants in Brazil mainly utilize sugarcane as the substrate, sugar beet is primarily utilized in the EU, and corn is the prime substrate for ethanol production in the US. The fermentation process comprises various consequently occurring chemical interactions, which change the carbohydrates found in the starch or sugar content of the organic species to carbon dioxide and ethanol. The fermentation process generally does not require high temperatures and is carried out at normal temperatures of nearly 25–37°C. Furthermore, the process may be conducted without the presence of oxygen molecules; hence, it is sometimes referred to as anaerobic digestion. Prior to reaching the fermentation step, an important step of milling exists. This can comprise dry or wet milling. Dry milling is generally known to entail a comparatively higher efficiency than wet milling. Furthermore, after the fermentation process, various types of processes may be conducted including distillation, adsorption, ozonation, and gas stripping. Distillation is a widely utilized technique in which the components of a mixture are separated based on their boiling points. For instance, compounds with lower boiling points evaporate from the mixture and the vapor can then be condensed to obtain the required product. Generally, in ethanol production plants the distillation process aims to separate water from the produced ethanol. One challenge associated with this technique is the presence of other volatile chemical species that can evaporate alongside ethanol, thus making it difficult to obtain a pure ethanol product output. Furthermore, as distillation is associated with a high operating cost owing to continuous boiling and condensation, other alternative methods to replace distillation are being investigated. Adsorption is one such method in which the chemical species in a given mixture are separated based on the adsorption properties over a given class of compounds. Another technique includes gas stripping, which is used to eradicate impurities from the ethanol mixture formed. Another recent technique introduced for ethanol production comprises biotechnological methods. In this technique, biomass generally obtained from waste can be utilized with suitable bacterial compounds to produce ethanol. The cellulosic and xylosic compounds are first separated from the waste biomass that is reacted with suitable bacteria for ethanol production.

3.2.2 Methanol fuel

Methanol has been tested extensively as an alternative fuel for gasoline. Specifically, in internal combustion engine-based machines and vehicles, methanol fuel has been utilized as an additive or complete fuel at various instances. However, its usage as a fuel has not flourished on a wider scale owing to the dominant gasoline fuel market. Nevertheless, in several racing cars, methanol has been used regularly due to safety issues. Methanol entails a lower cost as compared to ethanol; however, ethanol entails a lower toxicity as well as larger energy density. These fuels have been mainly focused for usage in internal combustion engines and their usage in fuel cells is also being investigated. Direct methanol fuel cells (DMFCs) include the usage of methanol fuel as an alternative to hydrogen in fuel cell technologies through electro-chemical interactions of methanol. The primary advantage of using methanol as a substitute for hydrogen is its higher volumetric density and safer properties making it easier to transport or store. However, the efficiency of hydrogen fuel cells still supersedes the efficiencies of alcohol fuel-based fuel cells. This is one of the reasons for utilizing methanol only as a hydrogen storage medium. Hydrogen can be stored chemically in the form of hydrogen and when fuel cell operation is required, methanol can be dissociated into its constituent compounds to obtain the required amount of hydrogen. This process is generally carried out via methanol steam reforming. It is thus commonly referred to as reformed methanol fuel cell. Although this type of fuel cell entails energy consumption during methanol to hydrogen conversion, the hydrogen fuel obtained provides much higher cell performances than DMFC.

In DMFC, the fuel input generally comprises a solution of methanol. The concentration of the solution can be varied according to the amount of fuel mass input required per unit time. The applications of the existing DMFC are limited to uses where not very high power requirements are needed. This is due to the limitations in the power output potentials associated with currently existing DMFC. Applications with low power needs such as small portable electrical devices including cameras or laptops can be operated with DMFC. Laptop chargers comprising DMFC and methanol fuel cartridges have already been developed and commercialized. Moreover, their usage in military machinery and equipment is also being investigated owing to their nearly noise free as well as benign operation. Although their power outputs are not as high as compared to hydrogen fuel cells, the amount of energy stored in a given volume of methanol can be higher than the energy stored in the same volume of hydrogen due to the low volumetric density of hydrogen.

The operation of an indirect methanol fuel cell is similar to that of hydrogen fuel cells in terms of the electrochemical reactions occurring in a hydrogen fuel cell. The primary difference is entailed in the dissociation or conversion of methanol to hydrogen, which is then fed to the fuel cell for power generation. However, the operation of DMFC varies from indirect methanol fuel cells. In DMFC, the methanol feed input reacts electrochemically in the presence of a catalyst at the anode. Also, oxygen is input at the cathode that completes the overall fuel cell reaction through its cathodic interaction. The overall fuel cell reaction occurring in a DMFC can be expressed as follows:

$$CH_3OH + 1.5O_2 \rightarrow 2H_2O + CO_2 \tag{3.29}$$

Further, the half-cell electrochemical reaction occurring at the cathode of the DMFC can be written as

$$1.5O_2 + 6H^+ + 6e \rightarrow 3H_2O \tag{3.30}$$

The corresponding anodic electrochemical reaction of methanol and water that emits electrons is written as follows:

$$CH_3OH + H_2O \rightarrow 6H^+ + CO_2 + 6e \tag{3.31}$$

As can be observed from the above half-cell reactions, these depict the DMFC operation with a proton conducting membrane. The membrane electrolyte in such a cell generally comprises Nafion materials that allow the passage of positively charged hydrogen (H^+) ions. However, similar to ethanol fuel cells, DMFC can also operate with alkaline or anion conducting electrolytes, which include the passage of negatively charged hydroxyl (OH^-) ions through the membrane electrolyte. The overall cell reaction remains the same as in the case of a proton exchange membrane electrolyte; however, the cathodic reaction for such cells can be expressed as

$$1.5O_2 + 3H_2O + 6e \rightarrow 6OH^- \tag{3.32}$$

As can be observed, oxygen molecules react electrochemically with water molecules to form negatively charged hydroxyl (OH^-) ions that are passed to the anodic side of the fuel cell where the following anodic half-cell electrochemical reaction is expected:

$$CH_3OH + 6OH^- \rightarrow 5H_2O + CO_2 + 6e \tag{3.33}$$

In either types of DMFC, the production of carbon dioxide is a disadvantage that increases the carbon footprints of such cells. Generally, in proton exchange membrane-based DMFCs, platinum catalyst is utilized to aid the electrochemical reactions. However, several types of catalysts have been developed and investigated in the recent past. The catalyst surface allows the methanol as well as water species to be adsorbed on the surface followed by subsequent interactions emitting electrons. Similarly, at the cathode oxygen species are adsorbed on the catalyst surface, which is followed by a consequent half-cell electrochemical reaction that completes the overall circuit. DMFC generally comprises a fuel input of a methanol solution with a molarity of nearly 1–3 M. Higher concentration solutions are not utilized owing to the higher possibility of fuel crossover in such solutions. In methanol fuel cells, methanol can pass from the anodic side of the cell to the cathodic side via diffusion through the membrane electrolyte. This crossover of fuel is aggravated when high concentrations of methanol solutions are used as fuels. However, lower concentrations of methanol fuel also pose limitations on the maximum amount of current that can be drawn from the cell. Lower concentrations of fuel comprise less number of methanol molecules that can react electrochemically and thus generate lower power through the series of half-cell reactions. Nevertheless, various types of methanol fuel cell systems have been developed that effectively manage the input as well as output of fuel. Moreover, another phenomenon that lowers the performance of DMFC comprises water drag. When the positively charged hydrogen ions are formed at the anode and pass through the membrane electrolyte to the cathode, a few of the water molecules are also dragged over to the cathodic side of the fuel cell.

When methanol reacts electrochemically, there are different pathways it may follow to complete the half-cell reaction. These may occur in parallel and comprise mainly the dehydrogenation of methanol as well as carbon specie oxidation along with stripping. The dehydrogenation pathway may first include the conversion of CH_3OH to $Pt-CH_2OH$ and the positively charged hydrogen ion (H^+) also referred to as proton. This is followed by multiple dehydrogenation steps where the $Pt-CH_2OH$ specie converts to Pt_2-CHOH and this specie further dehydrogenates until all of the hydrogen atoms are separated and converted into protons. At each step of dehydrogenation, electrons are emitted that generate the flow of current across the cell. There are several other types of dehydrogenation pathways suggested in the literature that are also possible in DMFC. Since adsorption of molecules plays a vital role in the fuel cell operation, several studies are focusing on developing suitable surface and crystal structures for platinum catalyst that provide better electrooxidation properties through enhancing the surface morphology as well as alloying with other metals. Moreover, during the series of adsorption processes there can be species of CO_{ads} formed, which are deteriorating for the platinum catalyst. Development of better types of catalysts for both anodic as well as cathodic reactions of DMFC has been dwelled upon extensively. As the catalyst plays a vital role in determining the performance of the fuel cell, enhancements in the electrochemical activity of methanol over any given catalyst can drastically improve the current DMFC technology. As far as the cathodic reaction of oxygen molecules is concerned, considerable progress has been made in developing new electro-catalysts that comprise nonnoble and inexpensive materials as well as entail higher electrochemical activity. However, better catalysts for the oxidation of methanol need to be developed. Various studies have been conducted on investigating new catalysts

Table 3.3 Summary of fuel cell performances of DMFC with corresponding operating parameters and system conditions.

Fuel concentration (M)	Power density (mW/cm^2)	Catalyst loading (mg/cm^2 PtRu)	Temperature (°C)	Catalyst
2	195	7	90	PtRu/NCNT-GHN
2	180	2	90	PtRu black
2	177	2	70	Pt-CoP/C
2	160	5	70	PtRu
2	137	3	70	PtRu/Vulcan
2	128	5	90	Pt/CNF
1	65	1	60	Pt-Ni$_2$P/C
4	44	2	25	PtRu/C
1	55	4	80	PtRu/BDDNP
2	105.2	2	50	PtRu-CoP/C

Source: From Ref. [17].

that comprise other metals, supports, particle size, etc. Table 3.3 summarizes the performances of different types of catalysts and DMFC investigations reported in the literature.

As Table 3.3 depicts, various types of catalysts are being investigated that provide varying performances. In addition, the molarity of methanol solution utilized was 1–2 M in most cases. The operating temperatures are also significant and have been found to enhance the fuel cell performances. Also, catalyst loading is another important cell parameter that can lower or improve the performance of the DMFC. However, as can be observed, most investigations utilized platinum-based catalysts. This shows the high dependency of fuel cell technologies on platinum, which is a noble metal and entails high costs. Some studies have investigated the integration of platinum catalyst with other nonnoble metals such as copper, nickel, and iron. Nevertheless, the fuel cell performances are not as promising as platinum and ruthenium catalysts. Some examples of catalysts free from platinum material tested for methanol electrooxidation include cobalt-copper alloy, zinc oxide on carbon, and graphene-based palladium-tin oxide. However, these types of electro-catalysts are suitable for usage in alkali mediums rather than acidic mediums such as the proton exchange membrane-based DMFC. The production of carbon dioxide during DMFC operation can result in the formation of carbonates, which can significantly deteriorate the fuel cell performance. Moreover, the reasons that necessitate the operation of non-platinum-based catalyst in alkaline mediums need to be further investigated. One of the reasons in case of a nickel-based catalyst has been suggested to be the formation of a layer of NiOOH, which forms when an alkaline potassium hydroxide medium is utilized. This layer activates the catalyst for further electrochemical reactions.

3.2.2.1 Production of methanol fuel

Methyl alcohol or methanol is a class of alcohol group of compounds that also entails various applications in several industries. It was initially referred to as wood alcohol owing to the production methodology that was utilized in those times. This methodology used wood to produce methanol through destructive distillation. In the polymer industry, formaldehyde is used extensively and the production of this chemical requires methanol. Hence, methanol is first used to produce other chemical compounds that are used in various industries. Methyl ether and acetic acid are also examples of chemical compounds that rely on methanol for their production. Moreover, another application of methanol comprises the production of gasoline, olefins, and hydrocarbons. Also, methanol is currently utilized in various countries as a fuel additive with gasoline. Nevertheless, the widespread usage of methanol as the primary fuel for internal combustion engines or fuel cells is expected to grow in the upcoming years.

The conventional production of methanol includes reaction of hydrogen with carbon monoxide. In this technique, natural gas reforming is performed through steam to produce synthetic gas that comprises both hydrogen as well as carbon monoxide along with carbon dioxide. These are the reactants required for methanol production. Hence, the produced synthetic gas is then passed to a methanol converter where the carbon dioxide reacts with hydrogen in the presence of copper and zinc oxide catalysts with alumina support to form methanol and water. The carbon dioxide that reacts with hydrogen can be obtained via water gas shift reaction that entails the reaction of carbon monoxide and water to form carbon dioxide and hydrogen. Hence, through a series of continuous steps and cycles, methanol is produced from methane gas. Similar to this method, methanol can also be produced through coal gasification. In this technique, the required synthetic gas is obtained via gasification of coal. New renewable and environmentally benign methods of methanol production are under investigation that are aiming to capture the carbon dioxide entailed in industrial flue gases and utilize it to produce methanol rather than reforming methane or gasifying coal. Another method of methanol production comprises enzyme-based conversion of methane to methanol. In this method, oxygenase enzymes are employed that convert methane in the presence of oxygen to methanol and water.

3.2.3 Propanol fuel

In the alcohol family of fuels, ethanol, methanol, propanol, and butanol have been considered for power generation. However, ethanol and methanol are more prominent in production as well as consumption. Specifically, for fuel cell operation, ethanol and methanol have been investigated primarily and a few marginal efforts have been directed toward studying the performance of propanol fuel cells. Several studies have been reported in the literature that focused on investigating the usage of 2-propanol fuel for direct propanol fuel cells (DPFCs). One such study conducted more than a decade earlier investigated a DPFC with a polymer electrolyte Nafion membrane and platinum with a ruthenium catalyst [18]. In this case also, an aqueous solution of 2-propanol was used at an operating temperature of 90°C. It was found that the fuel cell operates satisfactorily when low current densities of less than $200\,mA/cm^2$ are utilized. Also, it was reported that the efficiency of the DPFC is considerably higher than DMFC at low power densities of less than $128\,mW/cm^2$. Another study investigated a direct propanol fuel input embodied fuel cell with a split pH approach [19]. In this technique, the split pH cell comprised two streams of acidic as well as alkaline mediums. The acidic stream contained the propanol fuel along with sulfuric acid and hydrogen peroxide. The anodic catalyst comprised palladium black material and the cathodic catalyst entailed platinum black. The electrolyte used was composed of a cation exchange membrane. In addition to this, with the usage of a cathode that was comprised of carbon black, the peak power density was 101 and $115\,mW/cm^2$ for 2-propanol and 1-propanol, respectively. However, with the usage of the platinum black cathodic catalyst, the peak power density was reported to be 228 and $241\,mW/cm^2$ for 2-propanol and 1-propanol fuels, respectively. It was thus suggested that platinum is the most suitable catalyst for such types of fuel cells. Furthermore, another DPFC was investigated entailing a direct fuel input of 2-propanol [20]. The developed cell utilized platinum on carbon electrodes along with the liquid electrolyte. The maximum current was found at comparatively lower values of potentials. This was suggested to be due to an influence on the oxidation mechanism of propanol fuel. Moreover, a $22.3\,mW/mgPt$ of peak power density was reported and the majority of the polarization was found to be associated with the cathodic side of the fuel cell.

Table 3.4 lists the properties of the main types of alcohol fuels that have been considered for fuel cell applications. Primarily, interest has been shown in methanol and ethanol for utilizing as fuels in fuel cells. As discussed earlier, various investigations have been conducted and several commercial DMFC and DEFC have been developed in the recent past for different types of applications. Propanol has been studied as a fuel for fuel cells; however, not many promising outcomes have been achieved. Butanol, on the other hand, has been investigated primarily for internal combustion engines. As depicted in Table 3.4, the heating values, densities as well as percentage of hydrogen increase as the number of carbon atoms increases

Table 3.4 Summary of properties of different types of alcohol fuels mainly considered for fuel cell applications.

Alcohol	Chemical formula	Higher heating value (MJ/kg)	Density at STP (kg/m^3)	Hydrogen content (% by mass)
Methanol	CH_3OH	22.7	792	12.5
Ethanol	C_2H_5OH	29.7	789	13.0
Propanol	C_3H_7OH	33.6	803	13.3
Butanol	C_4H_9OH	33.1	810	13.5

in the alcohol group of compounds. However, owing to various other reasons such as production feasibility, cost, and fuel cell performance, ethanol and methanol have been the primary focus for usage as fuels in fuel cell technologies. Furthermore, as can be observed, the densities of alcohol fuels are suitably higher, especially, as compared to hydrogen, their volumetric densities are significantly higher. This is one of the primary reasons that these fuels have been considered to be good candidates for fuel cell technologies. Although the power outputs achieved from DMFC as well as DEFC are not as high as hydrogen fuel cells, the amount of stored energy per unit volume of fuel is substantially higher for both methanol as well as ethanol as compared to hydrogen fuel.

3.3 Alkane fuels

The alkane family of compounds comprises hydrocarbons that are extensively utilized in various industrial and commercial applications. The chemical formula of these compounds comprises $C_N H_{2N+2}$, where N denotes the number of carbon atoms and $2N+2$ denotes the corresponding number of hydrogen atoms. The widely used natural gas fuel comprises methane, which also belongs to this family of compounds. Methane is the basic compound of this group entailing one carbon atom and four hydrogen atoms, thus having the chemical formula CH_4. Further, the alkane with two carbon atoms and correspondingly having six hydrogen atoms is known as ethane. Moreover, the most commonly used transportation fuel of gasoline is also alkane based with 8 carbon atoms and 18 hydrogen atoms and is known as octane. In terms of uses in energy applications, alkanes have been predominantly utilized in internal combustion engines when the transportation sector is considered. In terms of stationary power applications, methane is used extensively across the globe in gas turbine-based power plants for power generation. However, in recent years, few alkanes have also been investigated as fuels for fuel cell technologies. The proceeding sections discuss these types of alkanes and their applications in fuel cells.

3.3.1 Methane fuel

Methane is the simplest alkane with only one carbon atom and forms the primary component of the extensively utilized natural gas. It is one of the crucial hydrocarbon fuels utilized in various countries throughout the world. One of the reasons for the widespread usage of natural gas and, thus, methane is the relative ease of availability and the comparative abundance in quantity. Being one of the natural resources that are always in high demand, natural gas is extracted from the earth through both onshore and offshore wells. Offshore wells comprise gas extraction from under the seabed and onshore wells entail extraction from underground resources. Natural gas that comprises mainly methane is used primarily in the applications of chemical feedstock as well as energy production. In the chemical industry, natural gas is extensively utilized to produce hydrogen through the SMR process. Although this method of hydrogen production entails considerable amounts of carbon emissions, it is currently the most well-established hydrogen production technique. However, new environmentally benign methods of hydrogen production are also being investigated, such as renewable energy-based water electrolysis and hybrid thermochemical hydrogen production cycles. This entails significant importance as in the ammonia production industry, an enormous amount of hydrogen is produced through the environmentally harmful method of SMR. Hence, to achieve a sustainable hydrogen and ammonia economy, it is essential to reduce the dependence on natural gas for hydrogen production. Apart from hydrogen production, natural gas is also used as a fuel for various energy applications. For instance, in kilns, residential water heaters as well as ovens, natural gas is the main fuel used across the globe. Moreover, in gas turbine power plants, natural gas is used extensively in various countries for power generation. In addition to this, several automobile manufacturers make natural gas-based vehicles that run on natural gas as the sole fuel for the internal combustion engine. Nevertheless, the usage of methane in fuel cell applications has been investigated recently for power generation. Specifically, high-temperature SOFCs have been investigated with methane fuels. Numerous studies have now been conducted and various systems have been developed based on methane fueled SOFC technologies. There are several advantages of electricity generation through electrochemical natural gas conversion. First, higher efficiencies are possible with fuel cells as compared to thermal power plants. For instance, methane fed SOFC power generation systems can entail efficiencies of more than 60%; however, natural gas-based thermal power plants generally entail efficiencies of nearly 30%–40%. This also means that less amounts of carbon dioxide emissions are the result for a given power generation quantity. A natural gas-based thermal power plant would result in higher carbon emissions per unit of power produced as compared to a natural gas-based SOFC. Moreover, methane fed SOFC entail an exhaust stream with a high amount of pure CO_2 and comparatively lower impurities of other chemical compounds that make it difficult to perform CO_2 capture. Unlike the combustion process of methane, an SOFC entails a membrane electrolyte that separates the air stream and the fuel stream. Hence, the nitrogen contained in the air stream is not allowed to pass to the fuel stream resulting in a concentrated stream of carbon dioxide in the exhaust, which is comparatively easier to handle for carbon capturing as compared to the exhaust gases of thermal power plants. In addition to this, as the operating temperatures of SOFC are generally not as high as the combustion chambers of thermal power plants, considerably lower amounts of nitrogen oxides are emitted. Nitrogen oxides are harmful environmental contaminants that are always undesirable due to various detrimental effects on both the environment as well as human health. In conventional combustion chambers of thermal power plants, the exhaust stream contains dilute carbon dioxide and is mixed with sulfur as well as nitrogen oxides. This poses several difficulties to capture the carbon emissions. Furthermore, methane fueled SOFC operate with minimal noise and entail a portable infrastructure that can be installed for both centralized and decentralized power generation. The SOFC technology also has the advantage of operating at high efficiencies even at low operating loads. However, the efficiency of gas turbines is reduced considerably when operating for lower loads. Also, SOFC-based power generation entails a comparatively less sophisticated infrastructure as compared to thermal power plants with less number of moving parts as well as energy consuming devices.

The methane fueled SOFC systems can be mainly categorized into three types of techniques. The first method consists of reforming methane fuel externally before it enters the fuel cell. In this method, SMR is performed outside the cell compartment in a separate reaction chamber that entails high temperatures and pressures along with catalysts. In addition, the produced syngas from the reforming reaction can also be reacted further via the water gas shift reaction to convert the carbon monoxide to carbon dioxide and more hydrogen. Hence, in this method, methane fuel is first converted into hydrogen that is then input to the SOFC for power generation. In this case, the electrochemical reactions entailed in the cell are similar to hydrogen fueled SOFC. The second method entails the reforming of methane fuel inside the cell compartment. In this method, methane along with water molecules is input at the anodic side of the SOFC, which is equipped with the required catalysts for SMR. The methane fuel is thus reformed inside the cell in the presence of catalysts and hydrogen is produced. The generated hydrogen then undergoes the anodic electrochemical reaction and the corresponding cathodic reaction of oxygen molecules remains the same. This method is possible owing to the high operating temperatures of SOFC as well as the release of energy as a virtue of electrochemical reactions, which allow the endothermic reforming reactions to take place within the cell. The release of energy from the exothermic electrochemical reactions is absorbed by the reforming reactions that are endothermic in nature. This aids in achieving high overall system efficiencies. Also, internal reforming aids in avoiding the requirement of a separate external unit and thus reduces the associated costs as well as sophistication. However, there are some disadvantages associated with internal reforming of methane in SOFC. The carbonaceous compounds formed at the anode can poison the catalyst layer or inhibit the catalyst activity. To reduce this occurrence, greater amounts of steam input are required to increase the ratio of steam to carbon. The third type of methane fueled SOFC includes the direct feeding of methane fuel to the cell without any internal or external SMR. This method helps to reduce the irreversibilities in the system integrated with internal or external reforming. Furthermore, direct methane fueled cells entail lower sophistications in the system design and are thus associated with lower costs as well as maintenance. The proposed half-cell electrochemical reaction of methane at the anode of direct methane SOFC is

$$CH_4 + 4O^{2-} \rightarrow CO_2 + 2H_2O + 8e \tag{3.34}$$

Also, the corresponding cathodic half-cell electrochemical reaction can be expressed as

$$2O_2 + 8e \rightarrow 4O^{2-} \tag{3.35}$$

The overall electrochemical reaction of the direct methane SOFC is thus similar to a complete combustion chemical reaction:

$$CH_4 + 2O_2 \rightarrow CO_2 + 2H_2O \quad\quad\quad (3.36)$$

These are the ideal proposed electrochemical interactions of methane; however, there are several associated complications. As can be observed from the anodic electrochemical reaction, water is formed in the reaction products. Hence, the steam reforming of methane fuel can occur due to the presence of water molecules. In the case of steam reforming of methane fuel at the anode, other chemical compounds including hydrogen and carbon monoxide can be formed that will affect the electrochemical reactions at both electrodes. Moreover, as each molecule of methane involves the transfer of eight electrons, the electrooxidation of methane is thus expected to take place in a series of elementary reaction steps. For instance, the initial step in the electrooxidation of methane would entail the bond activation of the carbon and hydrogen bond. This is followed by dehydrogenation where in a series of steps the hydrogen atoms are removed from the hydrocarbon molecules. Such interactions, however, are chemical reactions. Thus, the anodic reaction depicted in Eq. (3.34) in reality occurs through a combination of both electrochemical reactions as well as chemical interactions. Also, another distinguishing feature of direct methane SOFC includes the usage of pure methane fuel stream at the anode rather than a mixture of methane and other chemical compounds required for reforming. In case of external and internal reforming-type methane fuel cells, the fuel stream needs to be mixed with the required chemical species that are used during the reforming step.

The primary hindrance in the oxidation of the methane molecule is entailed in the energy barrier that needs to be overcome during the electrochemical or chemical interaction, which entails the highest barrier associated with the dissociation of the first carbon and hydrogen bond in the hydrocarbon molecule. This dissociation leads to the adsorption of the methyl group of species and then the dehydrogenation process. As conducting the oxidation of the methane molecule directly entails various challenges, the majority of methane fuel cells that have been developed and investigated considered the route of either the internal or external reforming of methane fuel. The syngas that is produced as a result of the reforming process is passed to the anodic compartment of the solid oxide cell. Moreover, as discussed earlier, in order to increase the amount of hydrogen at the anode and decrease the amount of carbon monoxide, the water gas shift reaction is also allowed to take place. Nevertheless, owing to the extra reforming step, both internal as well as external reforming-based methane fuel cells entail lower efficiencies than direct methane fuel cells theoretically. In addition to this, the theoretical change in entropies also varies for the reactions associated with these methods. For instance, the change in entropy associated with the oxidation reaction of methane Eq. (3.36) is considerably lower as compared to the hydrogen or carbon monoxide reactions. The latter include a higher amount of entropy losses and are known to be processes with large irreversibilities. Commercially built methane fuel cells entailing reforming-based SOFC technology have reported efficiencies of nearly 52%–60% considering the lower heating value of input fuel [12]. Furthermore, the main hindrances associated with natural gas fed fuel cells are the deposition of carbon species at the anodic side of the cell where the fuel or reformed gas is fed and poisoning of catalyst due to sulfur species. The carbon-based species are deposited at the electrode surface and inhibit the activity of the catalyst, which is also referred to as coking. The amount of deposition of carbon species is dependent on several factors and operating conditions. For example, the operating temperature of the SOFC affects the coking concentration and varying the temperature also changes the deposition of carbon species on the anode. Similarly, the steam to carbon ratio in reforming-based methane fuel cells also affects the amount of carbon species formed as well as deposited. Also, generally natural gas is utilized as fuel rather than pure methane as it is more readily available owing to the widespread usage in various industries. Naturally extracted natural gas includes small fractions of other chemical compounds that can include hydrogen sulfide or other sulfur-based species. In addition, to aid detection of leaks in pipelines, some smell entailing gases are added to hydrocarbon fuels, which can comprise sulfur-based chemical compounds. Sulfur and other chemical species entailing sulfur atoms are significantly deteriorating compounds for the electro-catalysts used in fuel cells. This remains one of the main challenges that needs to be addressed to flourish the development of methane fuel cells. Obtaining pure methane from natural gas would entail separate processes that would increase the overall costs of the system. Several research studies have been conducted in the recent past to investigate the sulfur poisoning mechanism in SOFC. Primarily, the poisoning of nickel-based catalysts via sulfur compounds, especially hydrogen sulfide, was the area of focus.

3.3.2 Ethane fuel

Ethane is the next compound in the series of alkanes that comprises two carbon atoms and correspondingly six hydrogen atoms. Conventionally, the primary hydrocarbon fuel such as natural gas that is extracted from the earth's crust is treated to obtain ethane or the refining of petrochemicals acts as a source of ethane production. The primary usage of ethane entails the production of ethylene. At ambient conditions of temperature and pressure, ethane exists in the gaseous form without any odor or color. Similar to other alkanes, ethane can be combusted to release energy that can be utilized for useful purposes. Generally, the natural gas obtained from underground wells can include ethane volume concentrations ranging from lower than 1% to higher than 6%. Several decades earlier, when natural gas was extracted, ethane was not separated from the obtained gas and the whole gas mixture was combusted for energy production. However, in the current petrochemical industry, ethane plays an important role as a feedstock for the production of other chemicals. Specifically, to produce ethylene, the steam cracking of ethane is performed. In this technique, heavy hydrocarbon molecules can be dissociated to form lighter molecules by steam dilution at high temperatures of more than 900°C. In addition to ethylene production, ethane is also being investigated to be utilized as a chemical feedstock for the production of other chemical compounds such as vinyl chloride and acetic acid. Moreover, cryogenic refrigeration is another application where ethane can be utilized as the working fluid.

However, the usage of ethane has also been investigated as a fuel for fuel cell technologies in various research studies. Specifically, the cogeneration of power and ethylene through ethane fuel cells has been an area of interest in the recent past. A study reported the development and performance assessment of a hybrid ethane fuel cell and ethylene production method. An SOFC-type cell was used with a proton conductive electrolyte comprising lanthanum-strontium and barium-cerium-based materials. At a temperature of 750°C, the power density of the ethane fuel cell was reported to be 320 mW/cm^2 [22]. Another study investigated the anodic catalyst comprising molybdenum carbide material for usage in ethane fed SOFC. The H$^+$ ion conducting electrolyte that was utilized was composed of BCZY material along with the molybdenum carbide electro-catalyst. The electrochemical assessment of the SOFC was found to be satisfactory with the new type of catalyst that showed good potential for usage in cogeneration systems entailing both ethane-based power and ethylene production [23]. Also, the dehydrogenation of ethane fuel was studied over a nano-size chromium oxide-based catalyst. This study also considered the route of power and ethylene cogeneration. The electrolyte, however, comprised a BCYN material that also entails H$^+$ ion conducting properties. It was found that the operating temperature significantly affects the output results. For instance, when the operating temperature was increased from 650 to 750°C, the power density was found to rise from 51 to 118 mW/cm^2. Also, correspondingly, the ethylene production was observed to rise from a yield of 8% to a yield of 31%. Importantly, the resistance to coking was investigated for the chromium oxide-based catalyst that showed better resistance than nickel or platinum-based catalysts [24]. Other y-doped barium and cerium oxide-based nano-powders were investigated for the cogeneration of ethylene and power through an ethane fuel cell. They were prepared through a combustion technique that provided perovskite phases. It was reported that the electrical conductivity as well as the carbon dioxide resistance both rise with the temperature of the sintering process. Moreover, an ethane fuel cell with an 800-μm thick electrolyte comprising BCY material was reported to provide a power density of 151 mW/cm^2 when platinum-based electrodes were utilized. In addition to this, the ethane conversion to produce ethylene was reported to be 35% when a 700°C operating temperature was employed [24].

3.3.3 Other alkane fuels

In addition to methane and ethane, other alkane hydrocarbon fuels have been investigated for fuel cell applications. These include several other alkane species such as propane, gasoline, diesel, etc. In this section, the research investigations conducted on these different types of alkane fuel cells will be discussed briefly.

3.3.3.1 Propane and butane fuels

The next chemical compound after ethane in the series of alkanes is propane. It contains three carbon atoms and correspondingly eight hydrogen atoms. Propane is already used as fuel in internal combustion engines, domestic heating, and in some countries for cooking. One main advantage of propane fuel comprises the ease of liquefaction as compared to natural gas. Propane can be liquefied and stored. This type of fuel is commonly referred to as liquefied petroleum gas. After becoming a liquid, as the volumetric density rises considerably, a larger amount of energy can be stored in relatively smaller spaces. On the other hand, natural gas or methane is comparatively harder to liquefy and is commonly stored as a compressed gas. Since high pressure gases are always hazardous, there are various safety issues associated with the storing of compressed natural gas. The usage of propane in fuel cells is not prominent and is currently undertaken primarily in research studies. In a recent study, a propane fed SOFC was investigated with fuel reforming. The propane fuel was converted and reformed into syngas via a membrane reactor. The produced syngas was then fed to the fuel cell. The electrolyte comprised oxygen permeation properties. The process of reforming was performed through the reaction with oxygen permeated and a ruthenium and nickel-based catalyst was utilized. They reported a hydrogen conversion of more than 90% as well as a syngas production rate of $22 \, mL/cm^2 min$ when utilizing a temperature of 850°C. In addition to this, the propane fed SOFC was found to fail when operated with unreformed fuel. However, the reformed fuel was found to provide stable fuel cell operation [26]. Furthermore, another recent study on propane fuel cells developed a propane fed tubular SOFC system, which was installed with a reformer providing partial oxidation. Also, a self-sustaining fuel cell operation was considered thermally. In the developed system, an SOFC stack of four tubular cells was utilized. Further, an after burner was also used to oxidize the exiting unreacted fuel and provide thermal energy to the fuel cell through inputs from both the anode and cathode of the stack. The output power of the four-cell SOFC stack was reported to be 12 W and the thermal performance was described to be an attainment of 780°C in an operation time of 1 h. The developed fuel cell stack with propane fueling system was thus described to be feasible for electricity production in portable applications [27]. Moreover, a research laboratory in Alaska had reported that tests were conducted successfully on propane fuel cells. A continuous operation exceeding 1100 h was performed and the efficiency was found to be relatively stable with no major deterioration. Furthermore, the propane fuel cell also provided heat as a useful output along with electrical power [28]. Another investigation focusing on the deposition of carbonaceous species during the process of dehydrogenation of propane molecules in fuel cells was conducted [29]. The membrane electrolyte comprised the proton conducting nature. In addition to power, the fuel cell also cogenerated propylene through the dehydrogenation of propane. This is similar to the cogeneration fuel cells developed using ethane for producing power and ethylene as discussed earlier. Both catalyst driven and electro-catalyst driven processes of dehydrogenation of propane were investigated and the deposition of carbonaceous compounds was compared for both types of processes. The catalyst used at the anode comprised chromium oxide material. They reported that the deposition of carbon reduces considerably when the current is drawn from the fuel cell. Also, potassium-based modification was made to the chromium oxide catalyst and the modified catalyst was found to provide higher resistance to the deposition or formation of carbon when no current is drawn from the fuel cell. However, the enhancement in the modified catalyst was not found to be considerable when the fuel cell was operated with a given current flow. Moreover, a propane-based fuel cell was developed and investigated that utilized propane flame for the fuel cell operation. The solid oxide-type cell was considered, which entailed an anode supported configuration. The propane flame provided dual functions of fuel reforming as well as thermal energy transfer to the cell for sustainable operating that does not require an external heating source to maintain the cell temperature. The developed cell showed high-performance characteristics and provided a power output density of $584 \, mW/cm^2$. In addition, the input flows of the fuel and oxidant mixture were varied, which was found to affect the performance of the fuel cell significantly [30]. Furthermore, to better assess the operation of propane fuel cells, a study was conducted on investigating the mechanistic reactions of propane during the fuel reforming process that occurs inside the cell, which is also referred to as internal reforming. The SOFC considered comprised an anode supported cell that entailed a nickel and yttria stabilized zirconia (YSZ)-based catalyst at the anode. The temperature of the cell was maintained at 800°C and the operation was investigated with an input propane concentration of 5%. Also, the propane-based SOFC was found to entail a stable operation. The output gases of the SOFC were reported to be hydrogen, carbon monoxide, methane, ethane as well as carbon dioxide. It was reported that a molar ratio of nearly 1.26 of carbon dioxide to propane associates the propane processing at a pseudo steady state condition [31]. Another propane fuel cell with a power output range of 150 W was investigated [32]. The SOFC-type cell was used and propane was employed as the fuel. A tubular type SOFC was developed that was installed with a partial reforming unit as well as heat exchangers and was reported to perform robustly, providing a promising portable power generator. The SOFC developed comprised micro cells in tubular configurations that were developed via processing procedures for ceramics. Also, the catalyst employed for partial oxidation of the fuel was made from cerium oxide and zirconium oxide materials that were supported on alumina-based supports. Various performance tests were conducted on the propane fuel cell with an integrated partial oxidation reformer and it was suggested that the developed system can be a suitable candidate for robust and portable power generation. Moreover, a propane fed SOFC was investigated with a silver-based anode [33]. A direct propane fed cell was considered where the fuel is directly input to the cell without the process of reforming. The propane fed fuel cell entailed an Ag-GDC anode and a power output density of $62 \, mW/cm^2$ was reported for the cell. In addition to this, the fuel cell was operated for 160 h to investigate the stability. This was performed at an operating temperature of 800°C. However, it was reported that a layer of carbon is formed on the silver-based anode when the cell is operated with propane fuel. Nevertheless, the rupturing of the formed layer of carbon was reported to provide more stability to the propane fed SOFC.

In addition to propane fuel cells, several investigations on butane fuel cells have also been conducted. In a recent study, a butane fed SOFC with a palladium-based anode was investigated. An operating temperature of 600°C was utilized in the study entailing direct input feed of butane fuel [34]. Also, the SOFC developed entailed thin-film-based configuration. They reported that the palladium utilized in the cell fabrication aided in the steam reforming as well as the water gas shift chemical reactions. Hence, as this results in more formation of hydrogen gas, palladium was concluded to provide superior fuel cell performance through enhancement of electrochemical activity as well as supply of hydrogen. In this study, palladium was considered as the secondary catalyst that was employed for aiding the butane reforming. It was installed in the cell through different techniques including infiltration and sputtering. Moreover, the cell configuration entailing palladium material at both the anodic support as well as functional layer was found to be associated with a comparatively lower deposition of carbon. Another study reported the development of a butane fuel cell with a solid oxide electrolyte [35]. The SOFC developed comprised an YSZ electrolyte and a tubular configuration was employed. Further, the cathodic material comprised porous platinum and the cell anode entailed nickel and YSZ material. The power output of the fuel cell was reported to be 0.14 W when operating on a butane gas bomb. In addition to this, both operating temperatures of 900 and 1000°C were tested and the higher temperature was found to be more favorable. Furthermore, another investigation on butane fuel cells was reported that considered the microtubular configuration [36]. This study also considered solid oxide electrolytes. The microtubular cells entailed diameters of lower than 2 mm. The study reported that the performance of anodes comprising nickel stabilized zirconia material was reduced considerably during an operation time of 3–4 h owing to deposition of carbon. This was reported to occur while the fuel cell was tested at an operating temperature of 610°C and the steam to carbon ratio was also in the lower range entailing a ratio of neatly 0.044. However, another type of anodic material was investigated comprising nickel and gadolinium-based material. In this cell, the deposition of carbon was found to reduce significantly and the cell was reported to be capable of operating continuously for greater than 24 h of operation time. In addition, copper has been investigated as an additional metal that can aid in the electrooxidation process in butane fed SOFC [37]. The study reported the development of a copper and iron bimetal material comprising anode. The developed cell was investigated with both hydrogen as well as butane as fuels. The preparation method was described to entail the wet impregnation onto a cerium oxide-YSZ-based porous matrix. The power output density of the fuel cell with butane fuel was reported to be 240 mW/cm^2. This was obtained when the developed cell was operated at a temperature of 800°C. Moreover, it was found that the anode conductivity is enhanced after it is exposed to *n*-butane. Also, the iron loading in the electro-catalyst was found to affect the fuel cell performance. An increase in the iron loading was found to enhance the performances. The carbon deposition was reported to be in the form of carbon fibers on the anode after operation with butane fuel.

3.3.3.2 Diesel and gasoline fuels for fuel cells

In addition to the alkane fuels discussed earlier that have been studied for fuel cell applications, especially for SOFC technologies, diesel and gasoline fuels have also been investigated. Although both diesel and gasoline fuels are widely used in internal combustion engines worldwide, some studies have tested their usage in fuel cell technologies. Specifically, SOFC technologies have been considered for this purpose. A diesel fueled SOFC was reported to be developed with a reformer [38]. The SOFC components comprised metal-based materials. The developed system entailed an autothermal reformer followed by post processing including a desulfurizer as well as post reformer. The reformate was analyzed and it was reported that low concentrations of diesel fuel are observed after reforming. However, the steam concentration was found to be high, which led to undesirably high oxygen concentrations at the anode, causing the oxidation of fuel cell components. The developed cell entailed an area of $2500\,mm^2$ and was operated satisfactorily for an operation time of 1000 h. The rate of cell degradation was also investigated and was reported to be nearly 4% for every 1000 h of fuel cell operation. Another external diesel reforming-based SOFC was investigated [39]. The formation of carbon was tested in the reforming-based cell. A photoacoustic-based testing method was employed that provided the amount of concentration of solid carbon. It was found that the formation of carbon depends on various factors including the ratio of oxygen to carbon atoms, utilization ratio of fuel as well as the fraction of recycling at the anode. Also, a considerable dependency was found between the concentrations of ethylene and the formed carbon. Moreover, they concluded that the temperature of the reformer most strongly affected the amount of carbon formation. In addition, an autothermal diesel reforming-based SOFC system was developed with anode side gas recycling [40]. The study found that recycling the exhaust gas of the anode into the reformer provides a dependable recycle stream and aids in achieving high overall efficiencies. The autothermal reformer comprised a rhodium-cerium oxide-zirconium oxide-based catalyst in a single-tube configuration. It was found that both air and water inputs entail significant parameters that affect the overall performance. Also, the reforming temperature and reactant space velocity were not found to affect the recycle process considerably. The optimal ratio of the recycling gas stream was reported to be 45% when a fuel utilization of 65% is being achieved.

In addition to this, gasoline fuel cells entailing solid oxide electrolyte-based cells have also been developed and investigated. In a recent study, a nickel and molybdenum-based catalyst entailing gasoline fueled SOFC was investigated [41]. The catalyst was reported to improve both the conversion of gasoline as well as hydrogen and carbon monoxide yields when operating at a temperature of 750°C. Nevertheless, it was found that button cells constrained the application of the catalyst owing to the large mechanical stresses arising from the restricted sintering. Hence, a tubular cell was developed that was found to entail higher catalyst application potentials. In addition, this aided in achieving higher hydrogen formation as well as cell lifetimes. The increase in cell lifetime was reported to be nearly 42% as compared to the button cell. Moreover, a gasoline fueled cell entailing the SOFC technology was developed entailing the molybdenum oxide-based anode [42]. The anode utilized was made through the electro-spraying methodology. The power output density obtained at the beginning of the fuel cell operation with gasoline fuel was found to be 30 mW/cm^2. This was found at a cell voltage of 0.47 V. Moreover, the open-circuit potential of the gasoline fueled cell was reported to be stable at a value of 0.92 V across an operation time of 1 day. Also, coking was not found to occur in the developed cell. Another nickel and molybdenum-based catalyst was investigated as a reforming layer in gasoline surrogate fueled SOFC [43]. The cell configuration comprised this internal micro-size catalyst layer for aiding in internal reforming of gasoline fuel. The gasoline surrogate of isooctane was investigated. Further, the SOFC entailed an anode supported configuration comprising nickel and YSZ-based materials. Also, an YSZ, ceria, and samarium-based electrolyte was utilized. In addition to this, the cathode used comprised a lanthanum-strontium-cobalt-iron material. The study found that the new type of catalyst developed aided in the enhancement of the electrooxidation through better reforming of the hydrocarbon into hydrogen and carbon monoxide. Further, it was found that the cell entailed a low polarization resistance at the new layer catalyst employed. The highest power output density obtained from the cell was reported to be 0.4 W/cm^2 while the developed cell was operated at a temperature of 750°C.

3.4 Ammonia

Owing to various favorable properties, ammonia is considered to be a promising alternative potential fuel for the upcoming future. First, it entails a competitive energy density of 4 kWh/kg providing a substantial advantage to be utilized as a fuel for power generation as well as transportation. Second, ammonia is comparatively easier to liquefy due to its high boiling point of −33.4°C at an ambient atmospheric pressure. Moreover, ammonia is considered to be a fuel that can provide solutions to the challenges associated with hydrogen fuel. The current challenges associated with hydrogen fuel discussed earlier, which hinder the flourishment of the hydrogen economy, can be addressed to some extent through the employment of ammonia fuel. Ammonia has a relatively high content of hydrogen of 17.7% by weight. In addition to this, ammonia entails a restricted range of flammability of about 16%–25% by volume when igniting in air. Table 3.5 depicts the comparison of physical properties of ammonia and hydrogen. At room temperature, hydrogen entails a considerably lower energy content per unit volume even though a high pressure of 10 MPa is used. However, one way of raising the density of hydrogen is through liquefaction. Specifically, for transportation as well as storage, liquid hydrogen is more preferable owing to a much larger volumetric density. Nevertheless, at atmospheric pressure,

Table 3.5 Comparison of physical properties of hydrogen and ammonia.

	Energy density (MJ/L)	Weight % H$_2$	Pressure (MPa)	Temperature (°C)	Density (g/L)
Hydrogen gas	0.9	100	10	25	7.7
Hydrogen liquid	8.6	100	0.1	−253	71.3
Ammonia liquid	12.9	17.7	1	25	603

very low cryogenic temperatures of nearly −253°C are needed to achieve hydrogen liquefaction. However, as depicted in Table 3.5, ammonia entails a significantly higher volumetric density as compared to both gaseous as well as liquid hydrogen. In addition to this, ammonia also entails a comparatively higher energy density when compared to both gaseous and liquid hydrogen fuel. Being associated with desirable characteristics such as lower flammability, carbon-free nature, cost-effectiveness, etc., ammonia is considered one of the most promising clean fuels that can play a vital role in the implementation of the hydrogen economy. Several challenges associated with the usage of hydrogen can be tackled through the usage of ammonia. For instance, the safety issues associated with the storage, transportation, and usage of hydrogen fuel will be significantly lowered with the utilization of ammonia fuel. Moreover, as it entails a higher volumetric density, a significantly higher amount of fuel can be stored in a given volume as compared to hydrogen. Further, being one of the most widely produced chemicals on a global scale, production of ammonia entails a well-established industrial infrastructure that can provide a stable supply of fuel. Currently, ammonia comprises one of the chemicals entailing the largest production globally. This is primarily attributed to its extensive usage in the fertilizer industry. In addition, it is also utilized as a feedstock for other chemicals and industrial processes. Ammonia production is estimated to require approximately 1.2% of the global primary energy. Also, it is estimated to result in nearly 1% of the total greenhouse gas emissions across the globe [44].

3.4.1 Ammonia production

There are various methods to produce ammonia. However, the most commonly employed technique globally comprises the Haber-Bosch process. This process involves reacting of hydrogen and nitrogen gases in the presence of an iron-based catalyst at high temperatures and pressures to produce ammonia. The primary detriment of this conventional methodology lies in the method of producing the required hydrogen. Approximately 70% of the global ammonia production relies on the usage of natural gas to produce hydrogen through the SMR process. In addition, nearly 20% of ammonia production is based on coal feedstock

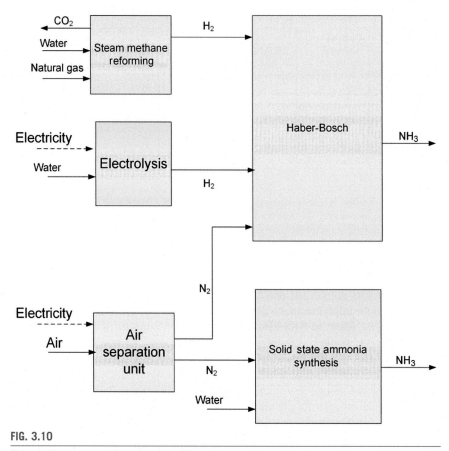

FIG. 3.10

Schematic representation showing different ammonia production methods.

[45,46]. Hence, these fossil fuel-based resources result in massive amounts of GHG emissions when utilized in ammonia production. Owing to this, several other environmentally benign methods of ammonia production are being investigated. Fig. 3.10 depicts the main types of ammonia production methodologies. These comprise primarily the Haber-Bosch process and the solid-state ammonia synthesis process. The first type of process is named after Carl Bosch and Fritz Haber who introduced the ammonia production method in the year 1913. In this process, hydrogen and nitrogen gases are allowed to react at nearly 400–500°C operating temperature and nearly 100–200 bar operating pressures. Also, an iron-based catalyst is utilized that aids in the production reaction. Moreover, the hydrogen required for the process is conventionally obtained via SMR. In this process, natural gas is allowed to react with steam at appropriate operating conditions and catalysts to obtain hydrogen. In addition, this process is associated with large amounts of CO_2 emissions. Hence, the environmentally benign route of water electrolysis is also

being considered for production of hydrogen. In this method, water and electricity are the required inputs and hydrogen and oxygen are the system outputs. In this way, a hydrogen production method that is free from carbon emissions can be employed. Several types of water electrolysis routes are being investigated including PEM electrolysis, alkaline electrolysis, solid oxide electrolysis, etc. Each of these methods entail its own characteristic properties and operating phenomenon. For instance, PEM electrolysis involves the passage of positively charged protons (H^+) through a membrane electrolyte. Also, alkaline electrolysis includes the passage of negatively charged hydroxyl (OH^-) ions through the electrolyte that can be an aqueous medium or an anion exchange membrane. Similarly, solid oxide electrolysis utilizes a solid oxide electrolyte and operates at high operating temperatures of nearly 800°C. With the usage of renewable energy to perform water electrolysis, a comprehensively benign hydrogen production method can be developed.

The air separation unit depicted is utilized to obtain pure nitrogen from air. This process involves cryogenic operations to condense and separate the required nitrogen for the ammonia synthesis process. Once the required nitrogen and hydrogen are obtained, they are reacted at suitable operating temperatures and pressures to form ammonia according to the reversible reaction:

$$N_2 + 3H_2 \leftrightarrow 2NH_3 \qquad (3.37)$$

The simplified process described above entails several other intermediate processes. A typical industrial ammonia synthesis process is depicted in Fig. 3.11. First, the feed to the process is treated to make it free of sulfur.

This process is important as sulfur is known to inhibit the activity of catalysts. Hence, to ensure the maximum product yield of ammonia, removing sulfur is an essential step. This is performed by passing hydrogen and allowing it to react with any sulfur compounds in the presence of catalysts to form hydrogen sulfide gas. Furthermore, the formed hydrogen sulfide is further allowed to get absorbed and react with zinc oxide to form zinc sulfide and water.

Next, the natural gas feed that is free from sulfur contents is allowed to pass to the SMR process. In this process, steam and natural gas react to form hydrogen and carbon monoxide. This process is followed by the water gas shift reaction where the carbon monoxide formed is allowed to react further with steam to form carbon dioxide and more amount of hydrogen. After one of these main system processes is performed, the carbon dioxide is separated in the CO_2 removal step. In this process, the pressure swing adsorption (PSA) or physical absorption of CO_2 in aqueous solutions is performed to separate the carbon containing compounds. Moreover, after this step, the methanation process is carried out to further eliminate the presence of any CO or CO_2 compounds. The pure stream of hydrogen produced is allowed to react with the traces of these compounds to form methane and water molecules. Lastly, once a pure and nearly carbon-free stream of hydrogen is obtained, the final step of ammonia synthesis is performed. In this step, the required nitrogen is obtained via cryogenic separation of air as discussed earlier.

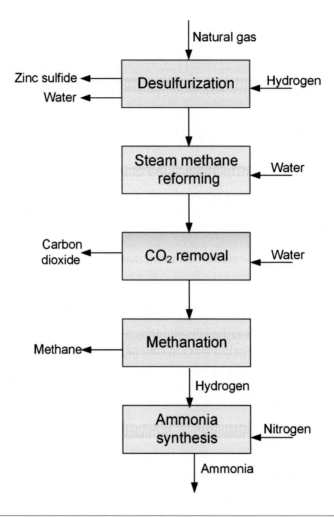

FIG. 3.11

Schematic representation showing various processes in an industrial ammonia production plant.

The solid-state ammonia synthesis process entails several differences as compared to the Haber-Bosch process. In this type of ammonia synthesis process, the input feeds to the reactor include nitrogen and water. The required nitrogen is obtained via a cryogen air separation unit as in a conventional ammonia production plant. However, instead of obtaining hydrogen through SMR or water electrolysis, direct feed of water is supplied to the reactor. In addition to the reactant feeds, electrical input is also supplied to the solid-state ammonia synthesis process. The overall reaction of ammonia production in such a system can be denoted as

$$6H_2O + 2N_2 \rightarrow 3O_2 + 4NH_3 \tag{3.38}$$

In contrast to the conventional chemical process for synthesizing ammonia, the solid-state process entails an electrochemical nature. Electricity is input to the reactor and through a series of electrochemical reactions occurring both at the anode as well as the cathode, ammonia is synthesized. The water molecules enter the reactor at the anodic side generally in the form of steam, where they are decomposed to form constituent hydrogen and oxygen species. Next, the hydrogen formed is adsorbed and electron stripping occurs. The positively charged hydrogen ions (H^+), also known as protons, are allowed to transmit through the solid electrolyte to reach the cathode. Here, electron acceptance occurs with hydrogen and nitrogen species combining to form ammonia molecules. Hence, this method eliminates the requirement of SMR for hydrogen production. Also, external water electrolyzers are not needed to produce hydrogen that consumes high amounts of electrical energy and entails both high operating as well as capital costs. Moreover, the necessity of operating at high pressures in the ammonia synthesis reactor is eliminated through the solid-state ammonia synthesis process. Further, this process can also be used to provide oxygen as a coproduct of the process. Several tubular bundle configurations are also possible with this method of ammonia synthesis. However, the proton conducting ceramic electrolytes are not easy to manufacture and entail their own production and manufacturing costs. Also, specific types of electro-catalysts are needed to be employed in such processes and efforts are being directed toward developing a new type of catalysts that can maximize both electronic as well as ionic conductivity, increase the surface area, and enhance the catalysis of reactions.

As a comparison, the solid-state ammonia synthesis process requires nearly 7000–8000 kWh to produce a ton of ammonia, whereas the natural gas-based process needs approximately 9700 kWh for the same amount of production. However, the water electrolysis-based process needs 12,000 kWh to produce the amount of hydrogen required to synthesize a ton of ammonia. Moreover, the capital costs of the solid-state process are expected to be comparatively lower owing to the lower sophistication in technology. Several efforts are being directed toward the development of green, renewable, and sustainable ammonia production methods. One such system is depicted in Fig. 3.12.

As depicted in the figure, this system utilizes renewable electricity to produce both nitrogen as well as hydrogen. The hydrogen is produced via water electrolysis and the required nitrogen is obtained through a PSA process. In this process, air and electricity are input to the system and pressure differentials are created in the presence of adsorption materials that selectively adsorb the nitrogen or oxygen at the differential pressure created. Once the adsorption process is completed, the pressures are again varied to release the adsorbed or the nonadsorbed gases.

For instance, in a nitrogen adsorption-based PSA system, the nitrogen molecules in air are adsorbed and the remaining nitrogen-free air is allowed to exit the system. When a sufficient amount of molecules are adsorbed, pressures are varied again to desorb or release these nitrogen molecules, thus allowing a pure nitrogen stream to be obtained. However, the purity of the nitrogen gas obtained depends on several factors including the adsorption material characteristics, number of cycles operated, etc.

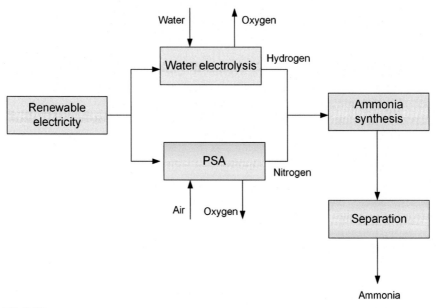

FIG. 3.12

Schematic representation showing a clean ammonia production method.

Nevertheless, the green ammonia production system provides an environmentally benign and sustainable method to synthesize ammonia from air and water. The key factor in the operation of such a system, however, depends on the reliability of the renewable electricity source. For instance, if solar or wind electricity is utilized to operate the system, the problem of intermittency will create hindrances in continuous and reliable operations of the system. During periods of low solar radiation intensities as well as wind velocities, sufficient energy may not be available to produce the required amount of hydrogen and nitrogen. Hence, energy storage units as well as backup energy sources will have to employed if such renewable energy resources are utilized. After the hydrogen and nitrogen gases are formed, they are allowed to react in the ammonia synthesis reactor. This reactor entails a similar operating procedure as in the case of the Haber-Bosch process. The produced ammonia along with unreacted nitrogen and hydrogen gases leaves the ammonia synthesis reactor and enters a separator and recycling loop. The separator is utilized to separate the ammonia from unreacted gases, which are passed back to the ammonia synthesis reactor after separation. This process of separation can be based on condensation of ammonia as in the case of conventional ammonia production plants, where ammonia is condensed and the liquid ammonia is separated from the remaining gases. Also, the adsorption-based ammonia separation method exists where the ammonia gas is adsorbed by appropriate selective adsorption agents and is thus separated from the unreacted gases.

3.4.2 Energy from ammonia

There are several ways of obtaining useful energy from ammonia. Fig. 3.13 depicts a schematic representation that summarizes several such types of ammonia energy routes. As depicted in the figure, it is possible to obtain energy from ammonia through conventional types of compression or spark ignition engines. However, the compression ratios required for ammonia fuel are different from those for diesel fuel. In addition, owing to the flammability limits, ammonia needs to be often blended with some combustion promoters in the form of conventional fuels to aid in the process. Furthermore, once satisfactory combustion of ammonia is achieved, ammonia gas turbines can be utilized. In such turbines, ammonia fuel is used instead of the conventional gas turbines that are fueled with natural gas as the fuel. Moreover, along with thermal engines, ammonia can also be used in electrochemical engines comprising fuel cell technologies, which is the primary focus of this book.

Along with power generation, ammonia can also be used in energy storage systems. This will be presented in the later sections of this book. When excess energy is available, it can be used to synthesize ammonia and when energy is required,

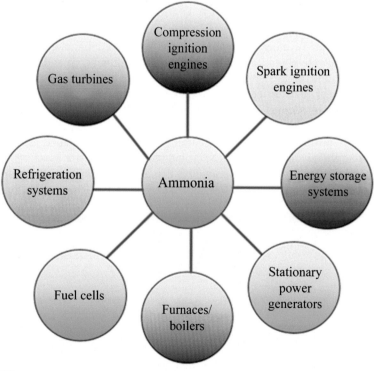

FIG. 3.13

Schematic representation of different ammonia energy systems.

the synthesized ammonia can be utilized for power generation through different types of technologies as discussed above. Hence, in this way ammonia acts as an energy storage medium. In addition to this, each ammonia molecule entails three hydrogen atoms. Thus, it can also be used as a chemical means of hydrogen storage. Also, other applications of ammonia fuel include the utilization of the heat of combustion directly in furnaces or boilers. Instead of burning carbon containing fossil fuels for heat generation, ammonia fuel can be combusted and the emitted thermal energy can be utilized for useful purposes. In the proceeding sections of this book, specific focus is given to ammonia fuel cells where electrical energy is obtained from ammonia fuel through a series of different electrochemical reactions.

3.5 Closing remarks

In this chapter, various different types of fuels are discussed that have been investigated for fuel cell technologies and applications. The comparative performances of each fuel type, their compatibility with different fuel cells, as well as their favorable advantages and nondesirable disadvantages are described. Also, the comparison of several physical properties of ammonia is made with other fuels showing the higher hydrogen content and volumetric density of ammonia that are favorable properties for fuel cell applications. Furthermore, the base energy density of ammonia is lower than that of other fuels; however, the reformed ammonia energy density entails a comparatively higher value. The high octane number as well as maximum compression ratio for ammonia fuel is also described in comparison with other fuels. Several types of fuels tested for fuel cell applications ranging from hydrogen to different types of alkanes as well as alcohols are discussed. The fuel cell performances of such systems are discussed in terms of the power output densities. Moreover, the production routes as well as current scenarios of these fuels are discussed considering the global conditions. Next, the usage of ammonia as an appropriate fuel for fuel cells is discussed providing details about its favorable properties. The conventional synthesis route is described along with the environmentally benign renewable energy-based route. Lastly, several applications of ammonia in different systems ranging from refrigeration to power generation are described.

Ammonia fuel cells

This chapter discusses ammonia fuel cells in detail with an emphasis on the different types of ammonia fuel cells that have been developed and their comparative performances. The classification categories for ammonia fuel cells are first described according to the type of electrolytes utilized and the method of fuel input used. The working principles of each type of ammonia fuel cell are described. Furthermore, the performances of ammonia fuel cells that have been developed in the recent past are analyzed based on the electrochemical catalysts used, operating temperatures deployed, and the type of electrolyte utilized.

Basic electrochemical parameters and equations are described in this chapter. The change in Gibbs energy for a given fuel cell reaction can be written as

$$\Delta G_{FC} = \Delta H_{FC} - T\Delta S_{FC} \tag{4.1}$$

where H and S denote enthalpy and entropy, respectively. After the evaluation of change in Gibbs energy, the reversible potential of the electrochemical cell can be found at ambient temperature (T_0) and pressure (P_0) as

$$E_{r,FC}^0 = -\frac{\Delta G_{FC,(T_0,P_0)}}{nF} \tag{4.2}$$

where the number of electrons is represented by n and the Faraday constant is denoted by F.

The actual fuel cell voltage under a given operating current density can be evaluated as

$$V_{FC} = E_{r,FC}^0 - V_{act,FC} - V_{ohm,FC} - V_{conc,FC} \tag{4.3}$$

where the terms subtracted from the open-circuit voltage denote the activation, Ohmic, and concentration polarization losses, respectively. The activation polarization loss occurring in the fuel cell at a given operating current density can be evaluated as

$$V_{act} = \frac{RT}{\alpha nF} \ln\left(\frac{J_{FC}}{J_{0,i,F}}\right) \tag{4.4}$$

where the operating current density is written as J_{FC}, the exchange current density is denoted by $J_{0,i,FC}$, the temperature is represented by T, the charge transfer coefficient is denoted by α, and the ideal gas constant is denoted by R.

Ammonia Fuel Cells. https://doi.org/10.1016/B978-0-12-822825-8.00004-9

In addition, the concentration polarization loss in voltage can be evaluated as

$$V_{con,i,FC} = \frac{RT}{nF} \ln\left(\frac{J_{L,i,FC}}{J_{L,i,FC} - J_{FC}}\right) \tag{4.5}$$

where the limiting current density is represented as $J_{L,i,FC}$ and the operating current density is denoted by J_{FC}. The limiting current density is calculated as a function of the diffusion coefficient (D), bulk concentration (C_B), and diffusion layer thickness (δ) according to

$$J_{L,i,FC} = nFD_{FC}\frac{C_B}{\delta} \tag{4.6}$$

Furthermore, the voltage loss due to Ohmic polarization in the electrochemical cell can be evaluated as

$$V_{ohm,FC} = J_{FC}R_{FC} \tag{4.7}$$

where the Ohmic resistances of the electrolytes and other cell components are denoted by $R_{FC/BT}$. Next, the operating current density of the fuel cell system is evaluated in terms of the Faraday's constant as

$$\dot{N}_{NH_3,FC} = \frac{J_{FC}}{nF} \tag{4.8}$$

The power output obtained from a given fuel cell can be evaluated in terms of the output voltage (V_{FC}), current (J_{FC}), and electrode area (A_{FC}) as follows:

$$\dot{W}_{FC} = V_{FC}J_{FC}A \tag{4.9}$$

4.1 Classification of ammonia fuel cells

Fuel cells are electrochemical devices that allow the production of electrical energy directly from the chemical energy of a fuel. When it comes to ammonia fuel cells, the primary classifications are made in accordance with the schematic representation shown in Fig. 4.1. First, the ammonia fuel cells are classified into two types: direct and indirect. The primary difference between the two types of fuel cell technologies lies in the type of feed that is input to the electrochemical cell. In direct ammonia fuel cells (DAFCs), the electrochemical cell receives an input of pure ammonia (NH_3) fuel.

However, in indirect ammonia fuel cells (IDAFCs), the ammonia molecule is first treated, reformed, or dissociated. The electrochemical cell in such type of fuel cells generally receives hydrogen gas formed from the dissociation of ammonia molecules as the fuel. The IDAFC entail an external dissociation unit that treats the ammonia to produce hydrogen fuel for the cell. Such systems utilize ammonia merely as a means of hydrogen storage. Owing to several challenges associated with hydrogen storage and transportation, ammonia is used to mitigate these hindrances by using it as a chemical means of hydrogen storage. Moreover, in such systems additional energy

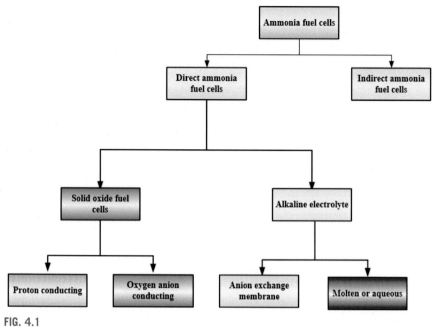

FIG. 4.1

Schematic representation of different types of ammonia fuel cells.

input is required to dissociate ammonia molecules that generally require high temperatures and are endothermic in nature. Hence, although power is generated from the hydrogen-fueled cell, a specific amount of energy needs to be supplied to the external ammonia dissociation unit to obtain a continuous supply of hydrogen. On the other hand, DAFC include a direct input of ammonia fuel without any external dissociation or decomposition of NH_3 molecules. As shown in Fig. 4.1, DAFC includes different types of fuel cells. These are primarily distinguished on the basis of the type of electrolyte utilized. The DAFCs are mainly classified into direct ammonia solid oxide fuel cells and alkaline fuel cells, depending on the type of the electrolyte used. The direct ammonia SOFC can be further classified into oxygen anion conducting or proton conducting fuel cells depending on the type of electrolyte operation. The oxygen anion conducting direct ammonia SOFC entails the transfer of negatively charged O^{2-} ions through ceramic electrolyte. However, the proton conducting DAFC includes the transfer of positively charged H^+ ions through solid oxide electrolyte. In addition to this, the direct ammonia alkaline fuel cell can also be classified as molten alkaline or anion exchange membrane electrolyte entailing DAFC. The primary difference in all the above-mentioned DAFC lies in the type of electrolyte employed. Different types of electrolytes dictate different electrochemical reactions occurring within the cell and hence different electrochemical operations are observed. In addition to the different electrolytes, these DAFCs also differ in their operating temperatures. The SOFC entailing DAFC include the usage of comparatively higher temperatures than alkaline DAFC. Also, the molten electrolyte-based cells include the utilization of temperatures above the melting point of alkaline salts such as sodium or potassium hydroxide. Hence, such fuel cells entail operating temperatures lower than SOFC but higher than cells comprising membrane electrolytes. The anion exchange membrane electrolyte-based DAFCs can operate at room temperature and do not require high temperatures unlike the SOFC or molten alkaline cells. However, owing to the comparatively lower operating temperature, the rate of electrochemical activity of ammonia molecules is also low, leading to lower power densities, voltages, and overall performances. In the proceeding sections, each type of DAFC is discussed in detail, and the relevant studies conducted on the development and enhancement of these technologies are also described.

4.2 Direct ammonia fuel cells entailing oxygen anion conducting solid oxide electrolyte

The DAFC operating with the transfer of negatively charged oxygen anions have been developed and investigated in the recent past. Ammonia is fed at the anodic side of the electrochemical cell that is maintained at temperatures in the range 500–1000°C. At this high temperature, the ammonia molecules are expected to dissociate internally into constituent hydrogen and nitrogen molecules, which can be depicted by the following equation:

$$NH_3 \leftrightarrow \frac{3}{2}H_2 + \frac{1}{2}N_2 \tag{4.10}$$

Furthermore, the cathode feed comprises generally pure oxygen or air. The oxygen molecules that enter the cathodic compartment interact electrochemically in the presence of a suitable electro-catalyst and accept electrons to form negatively charged oxygen anions. This half-cell electrochemical interaction at the cathode that is characteristic of such type of DAFC can be expressed as

$$\frac{1}{2}O_2 + 2e \rightarrow O^{2-} \tag{4.11}$$

As the oxygen molecules at the cathode and electrolyte interface keep converting to O^{2-} ions, the negatively charged ions keep migrating through the ceramic electrolyte to reach the anode and electrolyte interfaces. Here, the oxidation reaction or the half-cell anodic reaction of the fuel cell takes place. If all of the ammonia molecules are converted to hydrogen molecules owing to the high cell temperature, the anodic reaction can be expressed as

$$H_2 + O^{2-} \rightarrow H_2O + 2e \tag{4.12}$$

The fuel cell operation for the ammonia-fueled SOFC-O is depicted in Fig. 4.2. As depicted in the figure, the exhaust stream of the anode comprises unreacted ammonia and hydrogen along with the reaction products of nitrogen and water

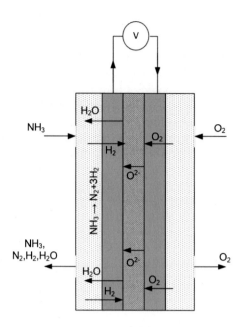

FIG. 4.2

Schematic representation of the operation of an ammonia-fueled SOFC-O.

molecules. Also, some studies have suggested that nitrogen oxides may also be formed due to the high temperature and presence of oxygen molecules. Under these conditions, nitrogen molecules can react with oxygen ions to form nitrogen oxides. These are environmentally harmful and their presence in SOFC-O operation raises several concerns about its environmental performance. However, further studies need to be conducted to confirm the presence, ratios, and amounts of nitrogen oxides that are formed during a typical ammonia-fueled SOFC-O. Moreover, the electrons are emitted at the interaction surface of hydrogen atoms and the negatively charged oxygen anions, which travel through an external circuit. This flow of electrons created through an external circuit enables electrons to reach the cathodic side of fuel cell where the oxygen molecules accept electrons and get reduced to negatively charged O^{2-} ions that migrate through the solid oxide electrolyte to reach the anodic side of the cell. Hence, a continuous flow of electrons and ions is generated as a continuous feed of fuel and oxidant is supplied to the cell.

The electrolytes primarily employed in ammonia-fueled SOFC-O comprise yttria-stabilized zirconia (YSZ) and samarium-doped ceria (SDC) materials. A study conducted on an ammonia-fueled SOFC-O showed that power density of 168.1 mW/cm^2 can be obtained with the usage of an SDC electrolyte with a thickness of 50 μm. Moreover, the developed cell entailed low-cost nickel material for the anode and a samarium-strontium-cobalt oxide-based cathode. The power density reported was for an operating temperature of 600°C. Nevertheless, the same cell showed better performance when operated with hydrogen fuel providing a maximum power density value that was nearly 23.1 mW/cm^2 higher. Also, the performances at other temperatures were studied and increasing operating temperatures were reported to provide higher fuel cell performances [47]. Furthermore, another study utilizing a low-cost nickel anode along with an SDC electrolyte was conducted with direct ammonia feed. The cathodic material was composed of barium-strontium-cobalt-iron oxide-based material. In this cell, however, the electrolyte utilized was five times thinner with a thickness of nearly 10 μm. The output power obtained from the cell per unit cell area was reported to be 1190 mW/cm^2. This was obtained when the cell temperature was maintained at 650°C. Also, the power density was found to increase with the utilization of hydrogen fuel. The increase was found to be significant with a value of nearly 682 mW/cm^2 [48]. In addition to this, another SDC electrolyte entailing direct ammonia SOFC was tested [49]. They also employed the cost-effective option of using a nickel oxide-based anode with an electrolyte thickness of 24 μm. Also, the cathode utilized in the study was a combination of samarium, strontium, and cobalt oxide materials. The fuel cell performance was not found to be comparatively higher than other cells developed and the power density was found to be nearly 467 mW/cm^2 when utilizing an operating temperature of 650°C and assessing at the maximum value. Moreover, other type of solid oxide electrolyte entailing the YSZ material was also investigated for direct ammonia solid oxide fuel cells. Nevertheless, the performance of these fuel cells was comparatively lower than the SDC-based cells. For instance, a power density of 202 mW/cm^2 was found when assessed at the maximum value for a direct ammonia-fueled cell entailing an YSZ electrolyte [50]. The thickness of the electrolyte used was 15 μm. The low-cost nickel material was employed for the anodic fuel cell side and a lanthanum-strontium-manganese oxide-based cathodic material was used. The reported fuel cell performance was for an operating temperature of 800°C. Although the electrolyte entailed a considerably less thickness, the fuel cell performance was observed to be significantly low as compared to SDC electrolyte-based cells. In addition to this, another similar study was conducted with the same type of electrolyte entailing double thickness [51]. Also, the anodic material was composed of nickel composition. Moreover, the cathodic material was composed of lanthanum, strontium, and manganese oxide as in the previously described study. However, the power density values obtained in this study were comparatively higher. For instance, a 299 mW/cm^2 of power density was reported to be obtained when the cell was operated at 750°C and this value was observed to increase when a higher operating temperature was employed. An increase of nearly 227 mW/cm^2 was reported when the operating temperature was increased by 100°C. Another direct ammonia-fueled SOFC with considerably higher electrolyte thickness was conducted [52]. The study also considered an YSZ electrolyte, however, significantly higher thickness of 400 μm was utilized. Nevertheless, nonsophisticated electrode materials were employed. The anode was composed of nickel oxide material and the cathode was composed of silver material. The power density when assessed at the maximum value was found to be 60 mW/cm^2. This was reported to be observed when the cell operating temperature was set at 800°C. The low power density obtained in this case can be attributed to the usage of a high-thickness electrolyte. The higher the electrolyte thickness, the higher the resistance to ionic transfer. This further leads to higher amounts of polarization losses as well as irreversibilities. Also, another thick electrolyte entailing direct ammonia SOFC-O was investigated [53]. The YSZ electrolyte utilized was 200 µm thick. Further, nickel entailing anode was utilized in this study as well and the cathodic material was also composed of lanthanum, strontium, and manganese. This study also found low power densities of nearly 88 mW/cm^2, although a high fuel cell operating temperature of 900°C was used. Furthermore, when the operating temperature was reduced by 200°C, the power density was found to decrease significantly by nearly 50 mW/cm^2. Hence, both operating temperatures and electrolyte thicknesses play a vital role in determining the performance of direct ammonia SOFC-O. At high temperatures, the rate of dissociation of ammonia molecules increases. Thus, as the number of ammonia molecules dissociating and forming hydrogen molecules increases, the electrochemical interaction at the anode becomes more favorable owing to higher number of reactive molecules that can participate in the electrochemical reactions providing using power outputs. Moreover, platinum has also been investigated as the anodic material [54]. The study utilized a platinum-based anode along with the BCG electrolyte. Also, the cathodic material was composed of silver entailing material. However, the power density results were not substantially better as compared to the nickel-based anode and lanthanum-strontium-based cathode containing fuel cell. When the fuel cell temperature was set at 800°C, the power output density was observed to be 50 mW/cm^2 when assessed at the peak value. Furthermore, when the operating temperature was increased by 200°C, the power output density was found to rise by a value of nearly 75 mW/cm^2.

The different types of direct ammonia-fueled SOFC-O developed and investigated have been summarized in Table 4.1. The type of electrolytes utilized and their corresponding thickness have also been listed as well as the type of anodic and cathodic materials employed. In addition, the performance of the developed fuel cell is described in terms of open-circuit voltage as well as maximum power density.

These type of electrolytes as well as electrode materials have also been extensively investigated for usage with hydrogen fuel. However, as discussed earlier, their application with ammonia fuel opens a new range of possibilities to utilize direct ammonia fuel electrochemically to convert its chemical energy into useful electrical energy. These type of fuel cells, however, necessitate high operating temperatures of nearly 500–1000°C. To achieve satisfactory performances, the operating temperatures in the upper range need to be utilized as temperatures in the lower range of about 500°C do not provide sufficient power outputs. However, in case of exothermic reactions, it is generally expected that the reactor in which the electrochemical reactions take place will entail the heat generated and gradually the external source of heat input can be eliminated. Nevertheless, in case of ammonia-fueled SOFC-O, further investigation is needed to analyze the amount of heat generation per unit time and the corresponding temperature increase and control. For instance, initially SOFC-O can be provided external heat to maintain the high temperatures required. However, as the fuel cell operation is proceeded and ammonia fuel is input, the increase in temperatures should be recorded with time. Correspondingly, the external heat input can be reduced in steps to achieve a stable operating temperature as required by the cell.

Table 4.1 Direct ammonia-fueled oxygen anion conducting solid oxide fuel cells.

Peak power density (mW/cm²)	Operating temperature (°C)	OCV (V)	Electrolyte thickness (µm)	Electrolyte	Electrodes	Source
65	500	0.9	50	SDC	Ni-SDC (anode) SSC-SDC (cathode)	[47]
168	600	0.88				
250	700	0.83				
167	550	0.795	10	SDC	NiO (anode) BSCF (cathode)	[48]
434	600	0.771				
1190	650	0.768				
60	800	1.22	400	YSZ	NiO-YSZ (anode) Ag (cathode)	[52]
202	800	1.06	15	YSZ	Ni-YSZ (anode) YSZ-LSM (cathode)	[50]
467	650	0.79	24	SDC	NiO-SDC (anode) SSC-SDC (cathode)	[49]
65	800	1.02	200	YSZ	Ni-YSZ (anode) LSM (cathode)	[53]
88	900	0.99				

4.3 Direct ammonia fuel cells entailing proton conducting solid oxide electrolyte

Proton conducting solid oxide electrolyte entailing DAFCs include the passage of positively charged protons (H^+) through the electrolyte. These types of fuel cells are also high-temperature fuel cells that involve the dissociation of ammonia molecules into constituent hydrogen and nitrogen molecules. The hydrogen molecules formed as a result of this dissociation reach the anode and electrolyte interface where they interact electrochemically in the presence of a catalyst to form positively charged protons. In this process, for every molecule of hydrogen oxidized, 2 mol of electrons are emitted at the anode. Moreover, at the cathode, oxygen or air is utilized as the oxidants. The oxygen molecules reach the interface of the electrolyte and the cathode where the protons arrive migrating from the anodic side and an electrochemical interaction takes place between the oxygen molecules and the positively charged protons to form water molecules. This interaction denotes the cathodic half-cell reaction that acts as electron acceptor. The fundamental operating principles of direct ammonia-fueled SOFC-H are depicted in Fig. 4.3.

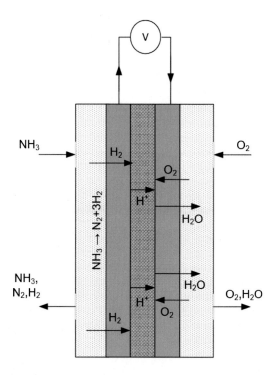

FIG. 4.3

Schematic representation of the operation of an ammonia-fueled SOFC-H.

The first step in the direct ammonia-fueled SOFC-H among a series of reactions is similar to that for SOFC-O where the ammonia molecules dissociate into hydrogen and nitrogen molecules. This reaction is depicted by Eq. (4.1). Next, as can be observed, the hydrogen molecules convert to positively charged protons at the anode and electrolyte interface. This anodic half-cell electrochemical reaction is depicted by the following equation:

$$H_2 \rightarrow 2H^+ + 2e \qquad (4.13)$$

Furthermore, the unreacted ammonia as well as the unreacted hydrogen and nitrogen molecules exit at the anode exhaust side. One of the advantages of ammonia-fueled SOFC-H is that it entails the absence of the formation of nitrogen oxides at the anode. Since the anode input stream generally comprises pure ammonia and no oxygen molecules are present, the nitrogen molecules do not obtain an opportunity to be oxidized. Moreover, at the cathode inlet stream, pure oxygen or air is input to the fuel cell. If air is utilized, the nitrogen molecules react with the protons coming from the anodic side of the cell to form other undesirable chemical species. However, if a pure oxygen input stream is used at the cathode inlet, these problems can be easily avoided. The oxygen molecules react electrochemically with the incoming protons to form water molecules, this half-cell electrochemical reaction occurring at the cathode can be denoted by

$$2H^+ + \frac{1}{2}O_2 + 2e \rightarrow H_2O \qquad (4.14)$$

As can be observed from the above equation, 2 mol of electrons are accepted for every 2 mol of positively charged protons that participate in the electrochemical reaction. In addition, the product of the half-cell electrochemical reaction includes the formation of water molecules where 1 mol of water molecules are formed for every 2 mol of electrons consumed. Hence, in this way, a continuous fuel cell operation is achieved, where the ammonia fuel is continuously fed at the anodic side of the cell and the proceeding reactions occur at the anode along with the corresponding cathodic reactions to generate a continuous flow of electrons. Some operating principles of ammonia-fueled SOFC-H are similar to that for SOFC-O. First, both entail the dissociation of ammonia molecules as an initial step owing to their similar high operating temperatures. Further, both result in the formation of water molecules where SOFC-O results in the formation of water molecules at the anode and SOFC-H results in their formation at the cathode. However, the primary difference between the two types of DAFCs lies in the ionic transfer in the electrolyte. The SOFC-O entails ionic transfer of negatively charged oxygen anions and SOFC-H includes the ionic transfer of positively charged protons. These dictate the differences in the electrochemical reactions between the two types of cell. In SOFC-O, as the anode side receives the negatively charged oxygen anions, the hydrogen molecules react with these anions according to the half-cell electrochemical reaction depicted by Eq. (4.5). On the other hand, in SOFC-H, the positively charged protons are formed at the anode that travels to cathodic side. Hence, the oxygen molecules are dictated to directly react with the incoming positively charged protons electrochemically in the presence of a catalyst.

Direct ammonia-fueled SOFC-H shows lower fuel cell performances than SOFC-O. The ammonia-fueled SOFC-H developed is mainly composed of barium-cerium-gadolinium (BCG)-based material. Furthermore, majority of the direct ammonia-fueled SOFC-H developed include the usage of platinum and nickel electrodes at the anode. One direct ammonia-fueled SOFC-H was developed utilizing platinum material at the anode as well as the cathode [54]. The electrolyte used was composed of a composite of barium-cerium-gadolinium-praseodymium (BCGP)-based material. Also, an electrolyte of thickness of 1300 μm was employed that can be attributed as one of the key reasons for the low performance observed. The fuel cell performance in terms of the power density assessed at the maximum value was found to be comparatively lower with a value of $35 \, mW/cm^2$. This was observed when the isothermal temperature was set at 700°C. The open-circuit voltage obtained for the cell was observed to be 0.85 V. In addition to this, another similar direct ammonia-fueled SOFC-H was developed [55]. The same type of electrodes comprising platinum material was employed at both the anode and the cathode. Also, the same electrolyte thickness (1300 μm) was utilized along with the same operating isothermal temperature (700°C). However, the electrolyte material was composed of BCG-based material rather than BCGP. The open-circuit voltage was found to be the same for both cells; however, the power output density was lower for the BCG electrolyte cell than that for BCGP. The power density was observed to be lowered by $10 \, mW/cm^2$ for the BCG-based cell as compared to the BCGP-based ammonia fuel cell. In addition to this, another BCG electrolyte-based DAFC was developed that provided a power output density of $32 \, mW/cm^2$ [56]. In this case, the electrolyte thickness of 1000 μm was employed. Also, similar operating temperature (700°C) was utilized. The open-circuit voltage, however, was observed to be 0.66 V, which is comparatively lower considering other similar cells developed and tested. The electrodes used in this study were also the same as described in the previous studies comprising platinum material. Although the same electrolyte was used with thickness lower than that in the studies described earlier, the power density was not observed to be higher. This can be attributed to various factors such as uncertainty, method of fabrication, repeatability, etc. The results from solid oxide fuel cells are not always repeatable and variations exist among different experimental tests. Moreover, another similar ammonia fuel cell entailing the usage of same type of platinum electrodes at either side with a BCGP electrolyte was fabricated [57]. The electrolyte thickness (1000 μm) used in the cell was similar to that used in the previous study. A power output density of $23 \, mW/cm^2$ was found for the ammonia fuel cell when the isothermal cell temperature was maintained at 600°C. Although the electrolyte thickness was same as in the previously described study, the power output was observed to be low owing to the lower operating temperature. The operating temperature plays a significant role in determining the performance of ammonia-fueled SOFC. Besides platinum material for the fuel cell electrodes, nickel-based materials were also investigated for the direct ammonia-fueled

SOFC-H. A nickel-barium-cerium-europium (Ni-BCE)-based anode and platinum-based cathode were utilized in a direct ammonia-fueled SOFC-H [56]. The electrolyte was composed of BCGP material entailing a thickness of 1000 μm. The open-circuit voltage was observed to be 0.92 V. Also, the DAFC was tested at temperatures 500, 550, and 600°C and the power output densities were found to increase with rising temperatures. For example, at an isothermal temperature of 500°C, the power density was reported to be 15 mW/cm^2 when assessed at the maximum value. This increased to 18 and 28 mW/cm^2 when operating temperatures were raised to 550 and 600°C, respectively. The results obtained hinted comparatively higher performances for DAFCs than only platinum entailing electrode cells. Another nickel-based electrode entailing direct ammonia-fueled SOFC-H was developed with a different type of cathodic material [58]. A lanthanum-strontium-cobalt oxide (LSCO)-based cathode was utilized for the fuel cell. The electrolyte material was composed of barium-cerium-gadolinium oxide (BCGO). Also, the developed cell included a low-thickness electrolyte that comprises a thickness of 50 μm. Further, varying isothermal operating temperatures were employed ranging from 600 to 750°C. The open-circuit voltages were found to decrease as the operating temperatures were raised. For instance, the open-circuit voltage at an operating temperature of 600°C was found to be 1.102 V. This decreased to 1.095 and 0.985 V as the isothermal temperatures were increased to 650 and 750°C, respectively. Moreover, the power output density assessed at the peak value was found to increase significantly with rising operating temperatures. At a temperature of 600°C, the power output density was reported to be 96 mW/cm^2. However, for a 100 and 150°C rise in isothermal temperature, the power output density was found to increase by 265 and 288 mW/cm^2, respectively. Another direct ammonia-fueled SOFC-H with the same type of BCGO electrolyte was developed with different electrode materials and electrolyte thickness [59]. An electrolyte thickness of 30 μm was employed. Also, the anode was composed of nickel-cerium-gadolinium oxide-based material. Further, the cathode is composed of barium-strontium-cerium-iron oxide (BSCFO)-based material. The developed fuel cell was also tested at varying isothermal temperatures and the open-circuit voltages showed a decreasing trend with increasing temperatures similar to that observed in the previously discussed study. The open-circuit voltage at 600°C was found to be 1.12 V, which decreased to 1.1 V when the operating temperature is raised by 50°C. The power density assessed at the peak value for the developed cell was found to be nearly 147 mW/cm^2 at an isothermal temperature of 600°C. Also, with an increase of 50°C in the operating temperature, the power output density increased by 53 mW/cm^2 providing a value of nearly 200 mW/cm^2. Another nickel entailing anode-based ammonia-fueled SOFC-H was developed [60]. In this new type of cell, anode was composed of nickel, barium, zirconium, cerium and yttrium oxide (BZCY) and the cathode was composed of barium, strontium, cerium, and iron oxide. This cell entailed the use of a different type of proton conducting electrolyte composed of the BZCY composition. Also, a comparatively lower thickness of 35 μm was employed for the electrolyte. Various different isothermal operation temperatures were investigated ranging from 450 to 750°C. The cell was observed to

Table 4.2 Direct ammonia-fueled proton conducting solid oxide fuel cells.

Peak power density (mW/cm²)	OCV (V)	Operating temperature (°C)	Electrolyte thickness (um)	Electrolyte	Electrodes	Source
35	0.85	700	1300	BCGP	Pt (Anode and cathode)	[54]
25	0.85	700	1300	BCG	Pt (Anode and cathode)	[55]
32	0.66	700	1000	BCG	Pt (Anode and cathode)	[57]
15		500	1000	BCGP	Ni-BCE (Anode) Pt (Cathode)	[57]
18		550				
28		600				
23	0.92	600	1000	BCGP	Pt (Anode and cathode)	[58]
96	1.102	600	50	BCGO	Ni-BCGO (anode) LSCO (cathode)	[58]
184	1.095	650				
355	0.995	700				
384	0.985	750				
147	1.12	600	30	BCGO	Ni-CGO (anode) BSCFO-CGO (cathode)	[59]
200	1.10	650				
25	0.95	450	35	BZCY	Ni-BZCY (anode) BSCF (cathode)	[60]
65		500				
130		550				
190		600				
275		650				
325		700				
390		750				
315	0.95	700	20	BCNO	NiO-BCNO (anode) LSCO (cathode)	[61]

provide a voltage of 0.95 V under no-load conditions. Further, the cell provided comparatively favorable performance, especially at high operating temperatures. Increasing isothermal temperatures were found to provide higher power output densities. For example, a low operating temperature of 450°C provided a power density of 25 mW/cm^2 when assessed at the peak value. However, as the operating temperatures were increased by 100, 200, and 300°C, the power densities were found to rise significantly by 105, 250, and 365 mW/cm^2. Thus, high operating temperatures as well as low electrolyte thicknesses have been found to be primary factors affecting the DAFC performances. Moreover, an anode composed of nickel oxide, barium, cerium, and neodymium oxide (BCNO) was utilized in a direct ammonia-fueled SOFC-H [61]. The electrolyte thickness employed in this cell was comparatively one of the lowest thickness reported for SOFC-H. It entailed a thickness of 20 μm and was operated at an isothermal operating temperature of 700°C. The no-load voltage was found to be of similar value (0.95 V) as reported in the study described earlier. Also, the power density assessed at the maximum value was also found to be similar. At the operating temperature of 700°C, the power output density was observed to be 315 mW/cm^2 at the maximum value. This was similar to the power density observed in the BZCY-based cell that provided a power output density of 325 mW/cm^2 at the maximum point. Although the cathodic material employed in this study was composed of LSCO-based composite, the results obtained were similar to that obtained in the earlier study conducted with the BZCY electrolyte and BSCF cathode. The similarity in results can be attributed to similar electrolyte properties owing to the presence of similar elements in the different composites utilized. The BZCY and BCNO ceramic oxides entail barium and cerium as common elements. A summary of the studies conducted on ammonia fueled proton conducting solid oxide fuel cells is given in Table 4.2.

4.4 Direct ammonia fuel cells with alkaline electrolytes

Alkaline fuel cell technology comprises one of the earliest developed fuel cell types that were also used in important applications such as in spacecraft. Increased interest surrounded alkaline fuel cells several decades earlier and they were also employed in other applications including vehicles and decentralized power generation. However, they were not sought after further owing to the introduction of various efficient and cost-effective options. Nevertheless, in the recent past scientists have again begun investigating this type of fuel cell technology owing to its various favorable characteristics. The primary advantage of alkaline fuel cells is entailed in the possibility of using nonnoble metal catalysts including nickel or iron at the fuel electrode, which can effectively reduce the cost of fuel cell technologies. Also, in alkaline fuel cells, cheaper cathodic catalysts such as silver can also be employed. These are attributed to the alkaline nature of the fuel cell which is associated with a high level of pH. At a high pH, the polarization losses are reduced as compared to acidic electrolyte cells such as PEM fuel cells. The oxygen reduction reaction that generally entails the highest irreversibilities in fuel cells is also facilitated in alkaline fuel cells owing to the high cell pH. Moreover, alkaline fuel cells comprise one of the most cost-effective fuel cell options. As compared to other types of fuel cells, they are much cheaper and several commercial alkaline fuel cells are already available. Although numerous efforts have been exerted toward the development and commercialization of hydrogen-fueled alkaline fuel cells, few direct ammonia-fueled alkaline fuel cells have also been developed. In DAFCs, ammonia is input at the cathode that reacts electrochemically with hydroxyl (OH^-) ions to emit electrons and form nitrogen and water molecules. This anodic half-cell reaction for a direct ammonia alkaline fuel cell can be expressed as

$$2NH_3 + 6OH^- \rightarrow N_2 + 6H_2O + 6e \qquad (4.15)$$

As can be observed from the equation, for every mole of ammonia that reacts electrochemically at the anode, 3 mol of electrons are emitted. Also, for every mole of ammonia that reacts at the anode, 3 mol of hydroxyl ions are needed. In a continuous operation fuel cell, these hydroxyl ions are generated at the cathode by the half-cell electrochemical reaction of oxygen and water molecules according to the following reaction:

$$\frac{3}{2}O_2 + 3H_2O + 6e \rightarrow 6OH^- \qquad (4.16)$$

The cathode inlet generally comprises a humidified air or oxygen feed that already contains water molecules, which can participate in the half-cell cathodic electrochemical reaction. Although water molecules are also formed through the anodic reaction, their travel to the cathodic side in alkaline membrane-based cells is not always favorable. However, in aqueous alkaline electrolyte-based cells, there is an abundant supply of water molecules that can participate in the cathodic half-cell reactions. The hydroxyl ions that are formed through this half-cell reaction at the cathode travel to the anodic side through a solid membrane, liquid aqueous, or molten electrolyte. The operation of an anion exchange membrane-based direct ammonia alkaline fuel cell is depicted in Fig. 4.4. As can be observed from the figure, the anode input comprises pure ammonia fuel and the ammonia molecules diffuse toward the interface of the anode and the electrolyte. At this interface, the hydroxyl ions arriving from the cathodic side interact with the ammonia molecules electrochemically to emit a stream of electrons that can travel through an external circuit. Furthermore, the anode exit stream consists of unreacted ammonia as well as the products of the half-cell electrochemical reaction including nitrogen and water molecules. At the cathode, correspondingly oxygen and water molecules enter the compartment and diffuse toward the cathode and electrolyte interface where the oxygen molecules react electrochemically with water molecules through electron acceptance to form hydroxyl ions. These negatively charged ions travel through the electrolyte to reach the anodic side of the cell and the fuel cell operation is thus continuously proceeded. The configuration depicted in Fig. 4.4, however,

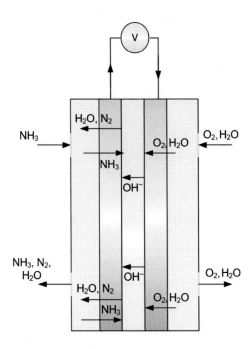

FIG. 4.4

Schematic representation of the operation of a direct ammonia-fueled anion exchange membrane fuel cell.

shows a anion exchange membrane-based cell. An aqueous electrolyte or molten electrolyte-based cells would entail a liquid or molten medium for ionic transfer. The electrodes along with the membrane electrode assembly (MEA) in such cells can be of different arrangements where the electrolyte is either continuously circulated or remains stationary between the two electrodes. In case of an electrolyte circulation configuration, an external pump would be utilized to provide the fluid flow. This would consume extra input power. However, circulating the electrolyte entails advantages that can aid in enhancing the fuel cell performances. One of the main advantages of electrolyte circulation include the prevention of concentration polarization at low current densities. The ions traveling in the electrolyte as well as the reactant gases accumulate in the vicinity of an electrode creating a hindrance to the diffusion of other ions, reactants, and product molecules. Circulating the electrolyte continuously during fuel cell operation aids in avoiding such circumstances at low operating currents, which might occur in the case of stationary electrolytes. The overall reaction for the direct ammonia-fueled alkaline fuel cell can be expressed as follows:

$$2NH_3 + \frac{3}{2}O_2 \rightarrow N_2 + 3H_2O \tag{4.17}$$

However, a key problem in alkaline fuel cells entails the reaction of carbon dioxide molecules with hydroxyl ions to from carbonates. Hence, CO_2-free streams of anodic and cathodic feeds need to be utilized. The carbonates formed owing to this interaction deteriorate the fuel cell performances. First, the number of hydroxyl ions reaching the reaction sites where ammonia oxidation occurs decreases. The reduction of these hydroxyl ions at the anode to participate in the actual electrochemical reaction decreases the voltage as well as the power density. Also, the ionic conductivity of an electrolyte reduces owing to the presence of carbonate ions. Specifically, in the case of aqueous or molten electrolytes, the carbonate compounds form precipitates that degrades the fuel cell performance considerably. However, this problem of precipitation is less serious in the case of membrane electrolytes. The chemical reaction representing this interaction of CO_2 molecules with hydroxyl ions is expressed as

$$CO_2 + 2OH^- \rightarrow CO_3^{2-} + H_2O \qquad (4.18)$$

Several direct ammonia alkaline fuel cells have been developed and investigated. The molten electrolyte-based cells include the usage of a molten salt comprising sodium or potassium hydroxide at temperatures in the range of 200–500°C. A direct ammonia-fueled molten alkaline fuel cell was observed to provide a power output density of $16\,mW/cm^2$ when assessed at the maximum value [62]. Nickel material was utilized for the electrodes. The maximum power density was found at an isothermal operating temperature of 200°C. Raising the temperatures to higher values, however, was found to provide higher fuel cell performances. For instance, increasing the temperature by 100, 200, and 250°C provided higher power output densities of 21, 31, and $40\,mW/cm^2$, respectively, at 300, 400, and 450°C. This can be attributed to the increase in the electrochemical activity at higher molten electrolyte temperatures. As the molten salt temperature rises, the hydroxyl anions attain higher energy that allows them to travel across the electrolyte at a faster rate. This allows faster reaction kinetics and thus better fuel cell performances. In addition to this, at higher temperatures, ammonia molecules may also dissociate into hydrogen molecules that participate in the electrochemical reactions. In such cases, higher temperatures increase the rate of dissociation producing higher number of hydrogen molecules that provide better fuel cell performances. Furthermore, at higher temperatures the polarization losses also reduce. As the ionic conductivity increases, the electrolyte resistance decreases. This reduces the Ohmic polarization losses occurring within the fuel cell. Faster reaction kinetics at higher temperatures aid in reducing the activation polarization losses. Higher diffusion coefficients of reactant and product chemical species aid in decreasing the concentration polarization, thus, leading to higher power output densities. Another DAFC with molten alkaline electrolyte was investigated [63]. In this study, platinum electrodes were utilized rather than nickel electrodes and the same type of molten sodium and potassium hydroxide electrolyte was employed. However, significant improvements were not found in the power output densities or no-load voltages with the usage of platinum electrodes. The no-load voltages were observed to be lower than the nickel electrode-based cell. For example, at 200°C the no-load voltage was 0.82 V in case of the nickel electrode entailing cell and 0.76 V in case of the platinum electrode-based cell. Also, at this temperature the power output density was observed to entail a comparatively lower value of 10.5 mW/cm^2 than the power output density (16 mW/cm^2) obtained in the nickel electrode-based cell. Nevertheless, the same trend of increasing power density with rising operating temperatures was found. As the cell isothermal temperature was increased from 200 to 220°C, the power density assessed at the maximum value increased from 10.5 to 16 mW/cm^2. The difference in the fuel cell performances observed in these two studies can be attributed to the variation in the type of fuel cell configurations. Moreover, in our recent study at the Clean Energy Research Laboratory (CERL) we also have investigated direct ammonia alkaline fuel cells with molten sodium and potassium hydroxide electrolyte [64]. The performance of the fuel cell was investigated in terms of the open-circuit voltage, power output density assessed at the maximum value, energy efficiency, and exergy efficiency. The isothermal operating temperatures of 220, 270, and 320°C were employed. The no-load or open-circuit voltage at a temperature of 220°C was observed to be 520 mV. This was found to decrease with rising cell temperatures. When a higher fuel cell temperature of 270°C was used, the no-load voltage dropped to 484 mV. Further, as the temperature was increased to 320°C, the voltage under no-load conditions dropped to 388 mV. This was attributed to the variations in the electrochemical reaction kinetics of ammonia oxidation with temperature in a sodium and potassium hydroxide medium. Investigation of the optimal range of temperatures was thus suggested. Moreover, the power density assessed at the peak value at the operating temperature of 220°C was observed to be 2.1 W/m^2. This entailed an increasing trend with rising cell temperatures. However, this rise was marginal owing to lower no-load voltages obtained at higher temperatures. For instance, at the higher isothermal temperature of 320°C, the power density at the maximum value was observed to be 2.3 W/m^2. The marginal increase could also be attributed to the rise in current densities with increasing cell temperatures. The short-circuit current density, for example, was found to rise from 14.5 to 22 A/m^2 as the operating temperature was raised from 220 to 320°C. This can be attributed to the reduction in the polarization losses with increasing electrolyte temperatures. At higher electrolyte and overall cell temperatures, the activation, Ohmic, and concentration polarization losses reduce. The activation polarization loss decreases owing to the higher ammonia conversion rates at higher temperatures. Further, the Ohmic polarization loss reduces owing to increase in the electrolyte conductivity and decrease in the electrolyte resistance. In direct ammonia molten electrolyte cells, the resistance of the electrolyte plays a vital role in determining the performance of the fuel cell. At lower electrolyte resistances, the ionic transfer and momentum within the electrolyte is enhanced and hence more number of reactant ions obtain an opportunity to react electrochemically. This can also be attributed to an increase in the diffusion coefficients of ionic species with temperature that are involved in the electrochemical interactions. Moreover, the concentration polarization loss is also affected by fuel cell temperature. As the temperature increases, the diffusion coefficients of both reactant and product molecules and ions increase. Hence, at higher temperatures the interacting species attain higher excitation as compared to lower temperatures. The higher excitation is directly related to the higher amount of kinetic energy these species entail. Thus, as the interacting species are more aggravated at higher temperatures, their movement in the vicinity of the electrodes is also enhanced. This aids in reducing the concentration polarization losses.

Apart from direct ammonia molten alkaline electrolyte fuel cells, anion exchange membrane-based DAFCs have also been developed and investigated. A direct ammonia-fueled anion exchange membrane-based cell was developed with chloroacetyl poly (2,6-dimethyl-1,4-phenylene oxide) dimethyl polyvinyl alcohol (CPPO-PVA)-based material [65]. Further, the anodic material was composed of chromium, decorated on nickel with carbon support (CDN/C). Also, the cathodic composition is composed of manganese oxide with carbon support (MnO_2/C). The developed DAFC was operated at room temperature and comparatively promising similar performances were observed as the molten electrolyte fuel cells. The open-circuit voltage, for instance, was found to be 0.85 V with ammonia fuel. Also, the power density assessed at the maximum value was observed to be 16 mW/cm^2. These results were similar to that obtained with the molten electrolyte cell with nickel electrodes operating at 200°C, where the no-load voltage was 0.82 V and the cell power density was 16 mW/cm^2. Although higher temperatures were utilized in the molten alkaline electrolyte entailing fuel cells, the similar results obtained can be attributed to various reasons. The distance between the electrodes plays a significant role in molten electrolyte cells. The higher the distance, the greater is the amount of resistance to ionic flow through the electrolyte, which results in higher polarization losses across the cell. However, in membrane-based cells, if the developed MEA entails a low overall thickness with membrane electrolytes entailing low thicknesses, then the resistance to ionic flow reduces considerably. Thus, it is essential to consider this trade-off in molten electrolyte and membrane-based cells. The molten electrolyte cells entail high temperatures which can be favorable for ammonia conversion, however, high amounts of polarization losses lower the fuel cell performances. In case of anion exchange membrane-based DAFCs, the low operating temperatures can be unfavorable for ammonia conversion. However, fuel cells entailing low membrane electrolyte thicknesses are associated with lower polarization losses that aid in enhancing the fuel cell performances. Moreover, the type of anion exchange membrane utilized also plays an important role. For instance, a different type of membrane from the previously study described was investigated for DAFCs [66]. The electrodes were composed of noble metal materials. The cathode was composed of platinum on carbon electro-catalyst and the anode was fabricated from platinum and ruthenium on carbon electro-catalyst. Although a higher operating temperature of 50°C was utilized, the no-load voltage was found to be lower than the previously discussed study. A no-load voltage of 0.42 V was obtained under these conditions, which is significantly lower than a no-load voltage of 0.85 V obtained in the previous study. In our recent study on DAFCs at CERL, anion exchange membrane-based cells were developed and investigated. Both single cell and 5-cell stack was developed and tested. Also, a new type of ammonia fuel cell system was developed that utilized aqueous ammonia as a medium to trap unreacted fuel and reuse it in the aqueous form. In addition to this, the performance of the developed fuel cell was investigated at different humidifier temperatures. The thickness of the membrane electrolyte was 500 μm that was composed of a quaternary ammonium functional group. Also, the polymer-based membrane was composed of polystyrene and divinylbenzene cross-linked polymer structure. Further, the electro-catalyst deployed comprised platinum black material. The fuel cell was designed to entail an MEA of the membrane electrolyte along with platinum black coated gas diffusion layers. The type of catalyst employed constituted 40% platinum on Vulcan material. Also, the catalyst loading comprised 0.45-mg platinum black per unit cm^2 of active area. The voltage under no-load conditions was observed to be 280 mV with gaseous fuel feed of ammonia at an operating humidifier temperature of 25°C. The humidifier temperature was increased by 45°C and 55°C and the fuel cell performance was investigated under these conditions. The no-load voltage was observed to increase to 300 and 335.5 mV at the humidifier temperatures of 60 and 80°C, respectively. Similarly, the power density observed at the maximum value was also found to increase correspondingly. The power output density at the humidifier temperature of 25°C, for example, was observed to be 6.4 W/m^2. This was observed to rise to 7.1 W/m^2 when a higher humidifier temperature of 80°C was utilized. This can be attributed to the enhancement in the electrochemical activity at the cathode due to the higher temperatures. As discussed earlier, higher temperatures aid in reducing various types of losses and irreversibilities occurring in the fuel cell. At the cathode, the reaction of oxygen with water molecules to form hydroxyl ions is enhanced owing to high temperature of water molecules entering the cathodic compartment. Moreover, the developed system entailed an unreacted fuel capturing subsystem that absorbed any unreacted ammonia in an aqueous solution. The formed aqueous ammonia solution was then investigated as fuel at varying solution temperatures. The temperatures were varied from a room temperature of 25°C to increased temperatures of 45 and 65°C. At room temperature, the fuel cell voltage under no-load conditions was observed to be 110 mV. However, when the solution temperature was increased to 65°C and the fuel cell performance was investigated, the voltage under no-load conditions was found to be 147 mV. Also, at a higher solution temperature, the rate of ammonia evaporation from the solution was found to rise. Hence, further higher temperatures could not be utilized. In addition to the open-circuit voltage, the power density evaluated at the maximum value was also found to rise with rising temperatures. At a solution temperature of 25°C, the power density was observed to be nearly 0.7 W/m^2. This was found to rise to nearly 1.3 and 2 W/m^2 at solution temperatures of 45 and 65°C, respectively. This can also be attributed to the decrease in polarization losses with temperature. As the solution temperature rises, the losses as well as the diffusion irreversibilities decrease leading to higher overall fuel cell performances. However, the power density values obtained for the aqueous fuel-based cell were found to be considerably lower. To achieve practical power outputs from an aqueous ammonia fuel-based cell, several improvements have to be made to the fuel flow, electro-catalysts, membrane, etc. For instance, during the experimental investigation, the aqueous ammonia fuel was found to crossover to the cathodic side of the membrane. This deteriorates the fuel cell performance significantly as firstly the oxidant molecules would not be able to reach the reaction sites effectively, reducing the number of electrochemical interactions and thus the overall performance of the cell. In addition, numerous molecules of ammonia fuel that were supposed to react electrochemically at the anode do not actually participate in the desired anodic reactions. This again reduces the electrochemical activity of the cell resulting in lower performances. Moreover, if the ammonia molecules start reacting electrochemically at the cathode, the electron emission and acceptance at the anode and the cathode would undergo undesired interactions. Further discussions about the DAFCs developed at CERL are entailed in the proceeding section. A summary of the various types of direct ammonia alkaline electrolyte-based cells developed and investigated is presented in Table 4.3.

Table 4.3 Direct ammonia-fueled alkaline electrolyte fuel cells.

Peak power density (mW/cm^2)	OCV (V)	Operating temperature (°C)	Electrolyte	Electrodes	Source
16	0.82	200	KOH +NaOH (eutectic mixture)	Nickel (both electrodes)	[62]
18	0.819	250			
21	0.817	300			
25	0.816	350			
31	0.813	400			
40	0.811	450			
10.5	0.76	200	KOH +NaOH (eutectic mixture)	Pt (both electrodes)	[63]
12	0.74	210			
16	0.73	220			
16	0.85	25	CPPO-PVA (membrane)	Anode: CDN/C Cathode: MnO$_2$/C	[65]
–	0.42	50	Anion exchange membrane	Cathode: Pt/C Anodes: Pt-Ru/C, Pt/C and Ru/C	[66]

4.5 Direct ammonia fuel cells developed at the Clean Energy Research Laboratory

At the CERL, DAFCs with alkaline electrolytes were developed and investigated. Both the anion exchange membrane-based cells and molten alkaline electrolyte-based cells were fabricated. In addition to this, a direct ammonia 5-cell stack was also developed entailing anion exchange membrane electrolytes along with other fuel cell components including flow channel plates, gas diffusion layers, etc. The exploded view of the developed single-cell anion exchange membrane-based cell is depicted in Fig. 4.5. In addition, the figure also describes the arrangement of different fuel cell components.

First, the end plates are depicted at the end of the cell on either side of the membrane. The end plates are utilized to enclose all fuel cell components and as can be observed from the figure, screws and nuts are used to tightly enclose the assembly. This is essential to prevent any leakage of gases from the cell during operation. Especially, as ammonia entails a pungent smell, leakage of reactant gases need to be avoided. Also, in hydrogen fuel cells, it is essential to ensure there is no leakage of fuel as it is highly flammable and can be very dangerous in applications such as transportation where the temperatures in the vicinity of the fuel cell can reach high values. Moreover, the end plates need be nonconductive to prevent short-circuit across the cell. Both end plates sandwich all the components and screws from the end plate on one side pass through the end plate on the other side and nuts tighten the sandwiched assembly.

Next, after the end plates, the flow channel or field plates are depicted. They are also known as bipolar plates when a stack is developed with multiple cells in series. They play an important role in allowing the reactant gases to effectively pass over the gas diffusion layers at both the anodic and the cathodic side of the cell. As can be observed from the figure, the flow field plates include grooved patterns that allow maximum exposure time for the reactants. The ammonia fuel is input at the anodic

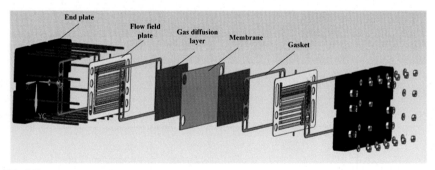

FIG. 4.5

Exploded view of the single-cell alkaline membrane direct ammonia fuel cell developed at CERL.

inlet and it passes through the grooves present on the flow channel plate where the molecules start diffusing to the gas diffusion layer. As the ammonia gas flows over the diffusion layer, the molecules keep diffusing into the layer such that near the anodic exit, the ammonia concentration is considerably lower than the anodic inlet. Hence, it is essential to design flow field plates that allow maximum exposure of reactant gases allowing higher amount of electrochemical interactions. However, as far as fuel cell stacks are concerned, the flow field plates need to be designed in such a way that they allow appropriate reactant flow over each electrode. In addition to allowing effective flow, the flow field plates are also utilized as current collectors in the present cell. The electrochemical measurements across the cell were made by connecting the positive and negative terminals across the flow field plates on either sides of the cell. Nevertheless, separate current collectors are generally employed in commercial fuel cell stacks that collect the current from multiples cells connected in series and are connected to the external circuit. Hence, the flow field plates are required to be electrically conductive in nature. In the present cell, stainless steel flow field plates are utilized that protect the plate from corroding or reacting with ammonia fuel. Ammonia is known to be corrosive chemical that can damage various materials through ammonia corrosion. Thus, it is essential to develop DAFCs that are resistant to ammonia corrosion and do not react under varying conditions. In case of corroding of fuel cell components by ammonia, the overall performance is affected adversely. The electrochemical activity across the cell reduces owing to other undesirable reactions occurring within the cell due to the presence of ammonia. The fuel cell power outputs as well as voltages are thus affected negatively in such cases. Moreover, stainless steel material also aids in applying high compressive forces on the cell or stack assembly. Compressive forces are applied through the end plates, screws, and nuts that enclose the overall assembly tightly. Higher assembly compression can be achieved if stainless steel plates are utilized. This will aid in reducing any leakage of reactants as well as decreasing the distance between the cell components. The distance between the fuel cell components also plays a role in determining the performances. Higher distances between cell components lead to greater cell resistances and thus lower the performances. On the other hand, if the distance between cell components is low, better fuel cell performances can be achieved. For instance, if a large distance exists between the gas diffusion layer and the membrane, higher resistance to specie diffusion will exist within the cell, which will lead to lower fuel cell performances and greater irreversibilities. Further, the flow field plates also act as bipolar plates in fuel cell stacks. The cells are arranged in a series configuration where the bipolar plate acts as the cathode for one cell and anode for the adjacent cell. In such fuel cell stacks, the fuel enters the anodic inlet and flows over only the anodic side of each adjacent cell and, on the other hand, the oxidant enters the cathodic inlet and flows only over the cathodic side of each cell. Thus, it is desirable to have bipolar flow channel plates that also entail higher electrical conductivities. After the flow channel plates, gaskets are depicted in the figure. Gaskets also play a vital role in fuel cell operation. Especially, when multiple cells are arranged in a stack, it is inevitable to develop a stack without gaskets. Gaskets also serve several purposes within the fuel cell. First, they prevent the reactants from leaking outside the cell through any gaps or passages. Commonly made from rubber material, the gaskets can be compressed to high pressures that enhance the sealing across the cell or stack. The higher the pressure applied, the lower the chances of gas leakages from the cell. Second, the rubber gaskets distribute the pressure applied evenly across the fuel cell stack. This is essential to ensure all cell components attain a uniform compression, which will lead to longer lifetimes owing to appropriate component configurations under high pressure. Next, the cell component after the gasket depicted in the figure is named as the gas diffusion layer. The gas diffusion layers allow the fuel as well as oxidant molecules to diffuse effectively toward the interface of the membrane electrolyte and the electrodes through embedded microporous layers. At the anode as fuel enters and flows through the flow field on the flow channel plate, the fuel molecules diffuse through the gas diffusion layer to reach the reaction sites. The layer in composed of appropriate passages that allow effective molecular diffusion. The gas diffusion layer utilized in the present cell comprises carbon material. Some characteristic properties of the diffusion layer are listed in Table 4.4.

In the present cell, catalyst-coated diffusion layers were utilized where the side facing the membrane was coated with platinum black catalyst (40% platinum on Vulcan). Hence, allowing the layer to provide both functions of gas diffusion and electrochemical reaction sites at the membrane and catalyst interface. The catalyst loading was $0.45\,mg/cm^2$. Furthermore, the thickness of the layer was 0.24 mm and it entailed a high porosity of 80% which provided sufficient space for gas diffusion. Further, paper-based diffusion layers were used in the present cell. However, cloth-based layers can also be utilized. The layers entail high compressive strength that can withstand the high pressures applied while sealing and tightening the assembly. Also, the diffusion layers are treated with PTFE material to provide a hydrophobic texture. This is important to prevent the cell from flooding with water during operation. At the anode, water molecules are produced continuously as the anodic electrochemical reactions take place. The molecules can flood the cell and in such cases, the performance deteriorates considerably. In case of water flooding, the reactant molecules are unable to reach the reaction sites effectively, which decreases the amount of electrochemical interactions. This can deteriorate the fuel

Table 4.4 Characteristic properties of catalyst-coated gas diffusion layer.

Polytetrafluoroethylene (PTFE) treatment	5% by weight
Porosity	80%
Plane electrical resistivity	$<12\,m\Omega\,cm^2$
Thickness	0.24 mm
Type of electro-catalyst	40% Platinum on Vulcan
Loading of platinum black	$0.45\,mg/cm^2$
Permeability for air	$1\pm0.6\,cm^3/(cm^2\,s)$

cell performance significantly. After the gas diffusion layer, the next fuel cell component depicted in Fig. 4.5 is the anion exchange membrane that plays a key role in the fuel cell operation. The membrane acts as the electrolyte for the cells and allows the passage of hydroxyl ions from the cathodic to the anodic side of the cell. The oxygen and water molecules are input to the anodic side of the cell where they react electrochemically to form hydroxyl ions. These ions need to be transferred to the anodic side of the cell where they interact with ammonia fuel electrochemically according to the half-cell anodic reaction of the direct ammonia alkaline fuel cell. Hence, the membrane electrolyte allows this transfer continuously and the fuel cell operation is proceeded. The membrane utilized in the present cell entails a strong base negative ion exchange functionality. Further, the membrane electrolyte comprises a quaternary ammonium function group. A schematic representation of the chemical structure of this functional group is shown in Fig. 4.6, where "R" denotes the group of atoms comprising alkyl or aryl functional group. Hence, the bond "N-R" represents the bond between these functional groups and the nitrogen atom. The positively charged quaternary ammonium ions entail the charge permanently and are not affected by the pH of the medium. In addition to this, the membrane utilized to develop the cell comprised chloride ionic form. Moreover, the thickness of the membrane was 500 μm, which is comparatively higher than other type of membranes utilized in DAFCs. This is one of the reasons why the developed cell entailed low fuel cell performances. It is recommended to develop low-thickness anion exchange membranes for DAFCs. This will aid in enhancing the performance of these type of fuel cells. As discussed earlier, as the membrane thickness decreases, the polarization losses also reduce. This can be attributed to the low distance the ions have to travel from the cathode to the anode.

The lower the distance between the interface of the membrane-anode and membrane-cathode reactive sites, the lower are the voltage losses and irreversibilities. The Ohmic losses reduce owing to the lower distance that the anions travel and lead to the overall reduction in cell resistance. Further, the current density that the membrane utilized could withstand was lower than $500 \, A/m^2$. Thus, this also limits the fuel cell output. As the current density is limited, the power output density

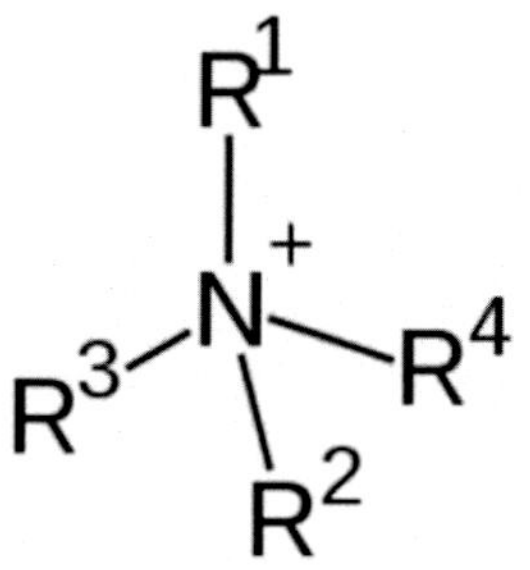

FIG. 4.6

Schematic representation depicting the quaternary ammonium ion.

Table 4.5 Characteristic properties of the anion exchange membrane.

Property	Value
Functional group	Quaternary ammonium
Functionality	Strong base anion exchange membrane
Ionic form	Chloride
Thickness (mm)	0.5 ± 0.025
Max. current density (A/m^2)	<500
Mullen burst test strength (psi)	>80
Maximum stable temperature (°C)	90
Color	Light yellow
Polymer structure	Cross-link between gel polystyrene and divinylbenzene

will also be lowered leading to low fuel cell performances. It is thus recommended to develop anion exchange membranes that can withstand high current densities to enhance the performance of current existing DAFCs. Further, the membrane electrolyte has a Mullen burst strength of more than 80 psi. Thus, it could be immersed in alkaline solutions for effective activation without bursting due to excess water absorption. The membrane polymer structure was composed of divinylbenzene and gel polystyrene cross-link structure. Before each experiment, the membrane has to be immersed in an alkaline solution of potassium hydroxide with a molarity of 1 M. These characteristic properties of the utilized anion exchange membrane are summarized in Table 4.5. There are other types of anion exchange membranes that have been developed in the recent past that entail lower thicknesses as well as lower resistances. Also, they entail hydroxide ionic form that would be more suitable for direct ammonia anion exchange membrane-based fuel cells.

A summary of the cell development procedure is depicted in Fig. 4.7. The figure depicts the stepwise procedure followed for developing the ammonia singe cell fuel cell as well as fuel cell stack. First, the membrane is prepared with gas flow channels on each corner corresponding to the anodic as well as cathodic inlets and exits to allow the passage of fuel as well as oxidant across the cell in a single-cell arrangement and across each adjacent cell in a fuel cell stack. Next, the catalyst-coated gas diffusion layers are placed on both the anodic and the cathodic side of the membrane. This assembly of membrane and diffusion layers can also be pressed with hydraulic presses to form an intact assembly. Also, they can be placed without hydraulic pressing as when the complete assembly is pressurized through end plates, the assembly undergoes compression. The next step in the procedure involving the manufacturing of flow channel plates. The stainless steel plates can be machined to form grooves of the desired shape and size. The serpentine flow field is commonly employed in these plates. Nevertheless, other types of flow fields can also be investigated. In addition, the inlets and outlets of the fuel cell at both the anodic and the cathodic side need to be precisely utilized to create effective fluid paths on the flow channel plates.

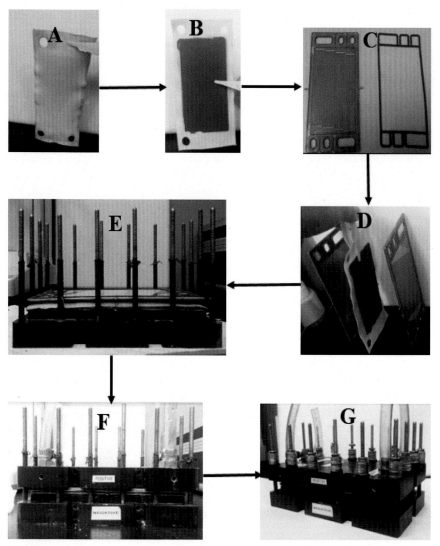

FIG. 4.7

Summary of the direct ammonia anion exchange membrane-based fuel cell development showing (A) anion exchange membrane, (B) catalyst-coated gas diffusion layers sandwiching the membrane, (C) flow channel plates with rubber gaskets, (D) membrane electrode assembly with flow channel plates, (E) membrane electrode assembly with back end plate, (F) front and back end plates, and (G) complete fuel cell stack assembly.

Further, in case of stack development, the dimensions and geometries of these inlet and outlet ports need to match for all plates utilized. This is particularly important to ensure that the fuel and oxidant flows effectively reach the respective anodic and cathodic sides of each cell in the stack. Along with the proper fabrication of flow channel plates, it is essential to make rubber gaskets and corresponding grooves on the plates. The plates cannot be assembled together for a fuel cell operation without the usage of gaskets as the gas molecules can leak out easily if the cells are not sealed sufficiently. Hence, to overcome this problem, grooves are precisely embedded on the flow channel plates and gaskets of the corresponding sizes are fabricated and placed on either side of the plate. After the flow channel plates are fabricated along with their gaskets, the assembly along with the membrane as well as gas diffusion layers is placed on the end plate. If one cell is to be investigated, only two flow channel plates are utilized. However, for developing a fuel cell stack, several plates need to be used according to the number of cells entailed in the stack. In case of stack development, the flow channel plate needs to be fabricated with two plates joined together along with gaskets placed in between. The flow channel plates act as bipolar plates in case of a stack arrangement. Hence, one side of the plate will act as the cathode of the first cell and the other side will act as the anode of the adjacent cell. Thus, the number of bipolar plates needed is one more than the number of membranes in the stack. For instance, if a 5-cell stack is developed, it contains five membranes and six plates. After placing the desired number of cells in series, the upper end plate is placed over the assembly as depicted in Fig. 4.7F. Both the upper and lower end plates sandwich the complete assembly in between them. Further, screws and nuts are used to tighten and compress all the fuel cell components against each other to minimize the distance between each component and to avoid any leakage of fuel or oxidant gases. This completes the fabrication of the anion exchange membrane-based DAFC. The performances of the developed cell and stack were investigated under different operating conditions. The experimental setups utilized for this purpose and the results obtained are discussed in the proceeding sections. Efforts need to be directed toward the development of new types of DAFC stacks that provide superior performances. The power output densities as well as current densities of the developed cell are not as competitive as hydrogen fuel cells. Hence, better electro-catalysts that are compatible with ammonia need to be investigated. However, although hydrogen fuel cells can provide higher power outputs, ammonia entails several other advantages that can overcome the challenges associated with hydrogen fuel. Thus, although the performances of DAFCs need to be improved and made competitive in comparison with hydrogen fuel cells, their performances do not have be matched.

4.5.1 Investigation of developed fuel cells

After developing alkaline fuel cells as described in the preceding section, their performances were investigated. The experimental setup utilized for this purpose is depicted in Figs. 4.8 and 4.9. Fig. 4.8 shows the actual photograph of the setup

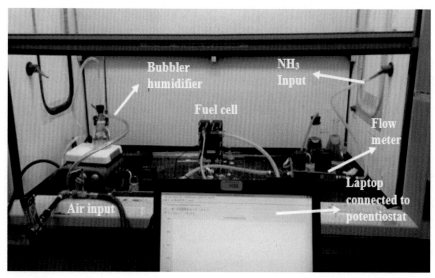

FIG. 4.8

Experimental setup to investigate the direct ammonia anion exchange membrane-based fuel cell.

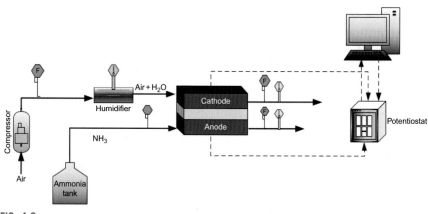

FIG. 4.9

Schematic representation of the experimental setup used to investigate the direct ammonia anion exchange membrane-based fuel cell.

and Fig. 4.9 depicts the schematic representation displaying all experimental components clearly. As can be observed from Fig. 4.9, the oxidant input utilized was humidified air. An air compressor was used to provide a continuous flow input of 1 L/min. This was used to provide an excess of stoichiometric oxygen molecules to achieve high utilization of fuel and low amount of unreacted fuel. Moreover, the air input from the compressor was passed through a bubbler humidifier to achieve humidification of the oxidant input to the cell. This step of humidification is required to provide the water molecules required at the cathode to initiate the half-cell cathodic electrochemical reactions. Also, the humidification keeps the membrane moist, which is essential for appropriate membrane electrolytic operation. Hence, the humidified air enters the cathodic compartment and the unreacted excess air exits the cell. Temperature sensors and flow meters are utilized to measure the temperatures as well as flow rates at the inlets and exits. Further, as can be observed from Fig. 4.9, the anodic side entails a pressurized ammonia tank that provides the fuel feed input. The anodic ammonia fuel input of 1 mg/s was utilized in the experimental investigation. Moreover, the inlet pressure was set at 1 bar for all experiments. Thus, ammonia fuel enters the cell or stack under these conditions and unreacted fuel exits the anodic compartment. The electrochemical measurements are made through a GAMRY Ref 3000 potentiostat. This device was suitable for the electrochemical measurements as the load and thus the current can be varied across a range of values and the output voltage can be recorded. From these measurements, the power outputs at different current densities can also be evaluated. Also, other electrochemical tests such as cyclic voltammetry, electrochemical impedance spectroscopy, chronopotentiometry, etc., can be performed.

The experimental test discussed above did not include the capture of unreacted fuel. From the system described earlier, another DAFC system was developed that included a reservoir for capturing unreacted fuel that formed aqueous ammonia. This was then investigated as fuel for the fuel cell. A schematic representation of this experimental setup is depicted in Fig. 4.10. In this system, the unreacted fuel exiting the anodic compartment is directed toward an aqueous ammonia reservoir that traps the incoming ammonia fuel.

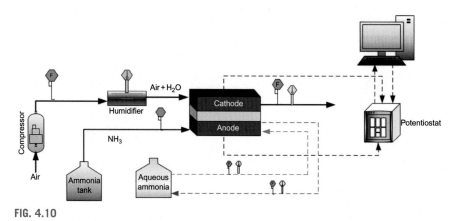

FIG. 4.10

Schematic representation of the experimental setup utilizing an aqueous ammonia reservoir for unreacted fuel.

Initially, a water reservoir is utilized that does not contain any ammonia. As the fuel cell operation begins and ammonia fuel is input to the cell at the anodic inlet, the unreacted ammonia fuel and some of the formed products exit the cell at the anodic exit. This stream is directed toward the reservoir where the unreacted ammonia is stored in the form of aqueous ammonia. In the experiments conducted, as the reservoir ammonia concentration reached 1 M, the fuel input was stalled and the fuel cell performance was investigated with aqueous ammonia fuel. In this way, a new system was proposed to store and utilize the unreacted ammonia fuel. Furthermore, the amount of unreacted ammonia at the anodic exit was observed to be considerably high and hence effective fuel management systems need to be investigated. Systems that can efficiently manage and reutilize the unreacted fuel will aid in avoiding the wastage of fuel and in achieving higher fuel utilization. The unreacted fuel could also be rerouted to the anodic inlet with a purge valve where the fuel entailed in the anodic exit stream is utilized at the anodic inlet stream with a certain amount purged outside. However, this method entails the disadvantage of fuel wastage, as some fuel has to be purged outside the system to avoid over-pressurization.

The system depicted in Fig. 4.10 entailing the concept of unreacted fuel capture, however, is also associated with certain challenges. The amount of ammonia gas that can be captured in a given volume of reservoir is limited by a saturation limit. Upon saturation of the aqueous ammonia solution, no further fuel can be absorbed or stored. Furthermore, the solution entailed in the reservoir is also associated with certain ammonia evaporation rates, which also result in fuel wastage. Nevertheless, both methods of rerouting the unreacted fuel to the anode inlet and capturing the unreacted fuel in a reservoir can be integrated to develop a more effective fuel management system. For example, the unreacted fuel at the anodic exit can be routed to the anodic inlet and the purge stream can be connected to the reservoir capturing the unreacted fuel.

The results obtained for the polarization behavior as well as the power output densities are depicted in Figs. 4.11 and 4.12. The three different lines represent the average results obtained for three different humidification temperatures investigated. To change the humidification temperatures, the temperature of the humidifier was changed to the desired value. The temperatures of 25, 60, and 80°C were used in the experimental investigation. Further higher temperatures could not be utilized as the anion exchange membrane entails a maximum stable temperature of 80°C. As can be observed from Fig. 4.11, the cell voltages under no-load conditions increase with rising humidifier temperatures. However, the rise is marginal as only the cathodic side of the cell is mainly affected by change in humidification temperatures. The voltage under no-load condition at a humidifier temperature of 25°C was observed to be 278 mV. However, as the humidifier temperature increased to 60 and 80°C, the no-load voltage was found to rise to 299.1 and 335.3 mV, respectively. These values are the average results of three experiments conducted at each humidification temperature. In addition to the no-load voltage, the current under short-circuit conditions was observed to increase with rising humidifier temperatures. For instance, the short-circuit current density was observed to be nearly

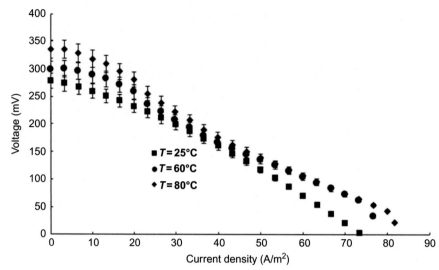

FIG. 4.11

Polarization results for the direct ammonia anion exchange membrane fuel cell with gaseous ammonia fuel at different humidification temperatures.

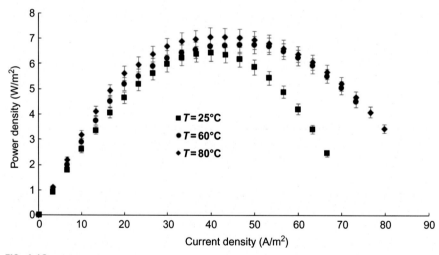

FIG. 4.12

Power output density results for the direct ammonia anion exchange membrane fuel cell with gaseous ammonia fuel at different humidification temperatures.

$73.3 \, A/m^2$ when an ambient humidifier temperature was utilized. Nevertheless, at a higher humidifier temperature of 60°C, the short-circuit current density increased to approximately $76.7 \, A/m^2$. Also, at a higher temperature of 80°C, the current was observed to reach short-circuit conditions at nearly $81.7 \, A/m^2$. These current values correspond to the average values evaluated for three experimental tests. The increase in no-load voltages with a rise in humidifier temperatures can be attributed to an increase in the electrochemical activity in the fuel cell. Especially, the cathodic half-cell electrochemical reaction will be enhanced owing to higher overall temperatures at the cathode. The water molecules entering the anodic compartment of the cell that react with oxygen molecules electrochemically, entail higher temperatures when the humidifier temperatures are raised. This aids in the oxygen reduction reaction at the cathode due to higher energy entailed in the interacting molecules. Moreover, as the humidifier temperature is increased, the anodic side of the cell is effected with time. As the fuel cell is continuously operated, heat is transferred from the higher temperature side of the cell to the lower temperature side. Furthermore, if the fuel cell is operated for longer periods of time, considerable heat generation within the cell occurs due to the electrochemical reactions. Hence, depending on the temperatures of the cell at the anodic and cathodic sides, heat transfer within the cell will occur from the higher to the lower side of the cell. The fuel cell behavior at varying humidifier temperatures in such cases needs to be investigated to determine whether increasing the humidification temperatures enhances or deteriorates the fuel cell performance. In case of excess heat within the cell, the temperatures can reach high values, which may damage the cell components. Specifically, the membrane entails a limit on the maximum operating temperature it can withstand. Nevertheless, commercial fuel cells include the passage of coolant through the cell continuously along with the reactant and oxidant flows. Water is commonly utilized as a coolant in these fuel cells, which is passed continuously across both sides of each cell to absorb the generated heat. Hence, providing appropriate cooling channels is essential for achieving satisfactory fuel cell performances when utilized in commercial applications. The bipolar plates discussed earlier play an important role in providing effective flow passages for the coolant across each cell. As mentioned earlier, two plates joined together can be employed as bipolar plates in stack arrangements. Passages are made within the two joint plates for the coolant to flow and absorb heat from each cell effectively. The coolant enters the stack and through the flow channels embedded between every bipolar plate, it flows in between each plate and absorbs the emitted heat from every cell. Further, this waste heat absorbed can also be utilized for other useful purposes such as cogeneration of electricity and heat. The operating fuel cell can provide electricity through the electrochemical reactions of the reactants and oxidants, whereas the waste heat absorbed by the coolant can be used to provide heating, thus, producing two useful outputs from the same system. The results of the power output density with varying current densities are depicted in Fig. 4.12. As can be observed from the figure, the power outputs also increase marginally with a rise in humidification temperatures, which is observed to occur around current density of $46.6 \, A/m^2$.

These are depicted in Figs. 4.13 and 4.14, respectively. Both efficiencies follow the same trend as the cell voltage. As the cell voltage decreases with increasing current densities, both exergetic and energetic efficiencies also decrease. The energy efficiency entailing the power output density at the maximum value is observed to

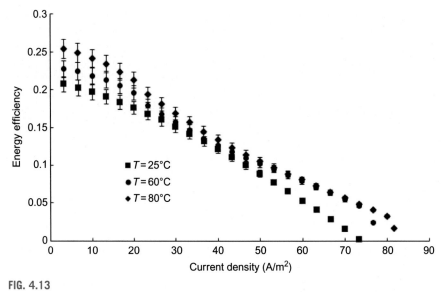

FIG. 4.13

Energy efficiency results for the single-cell anion exchange membrane fuel cell.

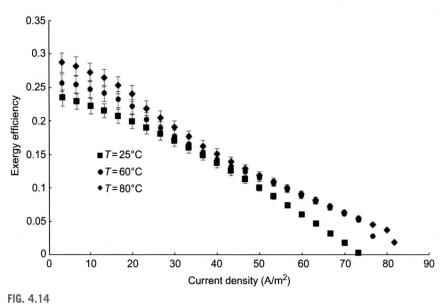

FIG. 4.14

Exergy efficiency results for the single-cell anion exchange membrane fuel cell.

be 12.1%, whereas the exergy efficiency is evaluated to be 13.8% when the humidifier temperature was under ambient conditions. Also, the energy efficiencies at the maximum levels of power output densities at higher humidifier temperatures are found to be nearly similar. As can be observed from the figures, both energetic and exergetic efficiencies entail similar values around the maximum power densities and entail higher differences in efficiencies at lower or higher current densities. For instance, both efficiencies are observed to be higher when higher humidification temperatures are utilized before the peak power density range.

However, after this range, the efficiencies at 60 and 80°C are found to be closely similar to each other. Thus, it is essential to investigate the optimal performances of a fuel cell during its design and utilization phases to ensure that the operation takes place in the vicinity of both maximum power densities and optimal efficiencies. Efficiencies decrease with increasing current densities owing to an increase in the polarization losses. As these losses increase, higher amount of useful energy is lost and higher irreversibilities are associated with fuel cell, leading to both lower energy and exergy efficiency. However, the power density entails a different trend. As the current density rises, the power density also increases up to a maximum point, which is referred to as the peak or maximum power density. After this value, the power output starts to decrease. This is attributed mainly to the polarization losses occurring within the cell. After the maximum value of power density, the polarization losses attain considerably high values such that an increase in current leads to a significant drop in voltage. This also results in a drop in power density.

The aqueous ammonia fuel developed through the unreacted fuel as depicted in Fig. 4.10 was also tested for the single-cell ammonia fuel cell. The polarization results for the cell with aqueous ammonia are depicted in Fig. 4.15. As can be observed from the figure, considerable increase in the fuel cell performance was observed with increasing ammonia solution temperatures. At an ambient solution temperature of 25°C, the voltage under no-load conditions was observed to be 110.1 mV. Also, the current density was found to entail a short-circuit value of nearly 20 A/m^2.

However, at higher solution temperatures of 45 and 65°C, the voltages under load conditions were found to rise to 130.1 and 147.1 mV, respectively. Also, the currents assessed at the short-circuit values are observed to rise to higher values at higher solution temperatures. These rises in voltage and currents can be attributed to the rise in electrochemical activity as well as rate of ammonia evaporation with temperature. At higher solution temperatures, the molecules entail more energy and thus the reaction kinetics are enhanced.

Also, at higher solution temperatures, the rate of ammonia evaporation rises due to the higher amount of energy entailed in the ammonia molecules. Further, when higher number of ammonia molecules evaporate from the solution per unit time, higher diffusion of molecules through the gas diffusion layers is expected. This results in higher number of ammonia molecules reaching the active sites, thus enhancing the fuel cell performances. Moreover, the power output density results for aqueous ammonia-fed single cell are depicted in Fig. 4.16. The power density values are observed to increase considerably with solution temperatures. At an

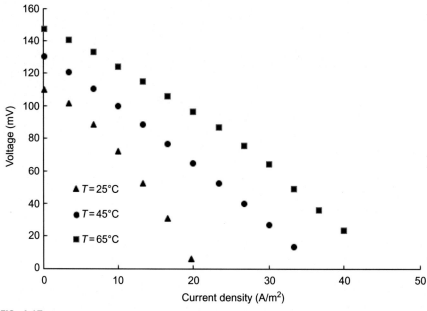

FIG. 4.15

Polarization results for the single-cell anion exchange membrane fuel cell fueled with aqueous ammonia.

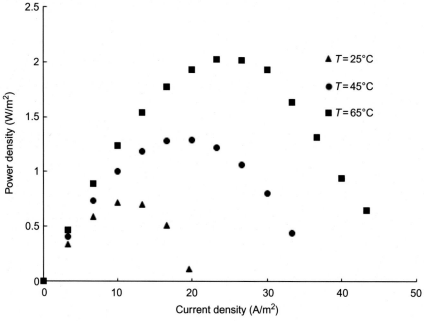

FIG. 4.16

Power density results for the single-cell anion exchange membrane fuel cell fueled with aqueous ammonia.

ambient solution temperature of 25°C, the power density assessed at the peak value was found to be 0.72 W/m². This was found to enhance considerably with rising temperatures. For instance, at a higher ammonia solution fuel temperature of 45°C, the power density assessed at the peak value was found to entail an increase of 0.57 W/m² as compared to ambient conditions. Also, at a higher solution temperature of 65°C, the power density observed at the peak value was 2.02 W/m². The increase in the power densities with a rise in the solution temperatures can be attributed to the excitation of ammonia molecules at higher temperatures. As discussed earlier, both the rate of ammonia evaporation and electrochemical activity increase with rising temperatures. This aids in achieving higher voltages, current densities, and thus power output densities. Although aqueous ammonia fuel provides output voltages as well as power densities, these are observed to be considerably lower than other types of fuel cells. Hence, it is recommended to utilize waste heat to extract the absorbed ammonia molecules from the aqueous trap solution. This can aid in providing a gaseous fuel input to the cell rather than an aqueous input and thus the fuel cell performances obtained for the gaseous fuel can be expected. After the investigation of a single-cell DAFC composed of an anion exchange membrane electrolyte, a 5-cell stack was also developed and its performance was assessed. The polarization curves for the 5-cell stack fed with gaseous ammonia fuel is depicted in Fig. 4.17. Also, the power output density results evaluated for the 5-cell stack fed with gaseous fuel are depicted in Fig. 4.18.

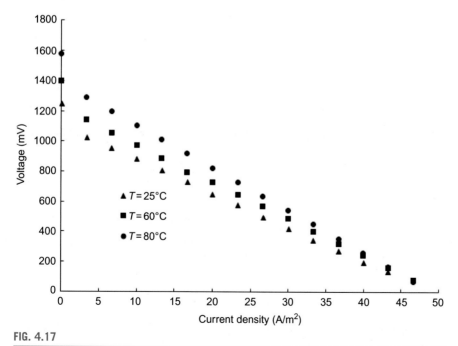

FIG. 4.17

Polarization results for the 5-cell anion exchange membrane fuel cell stack fueled with gaseous ammonia.

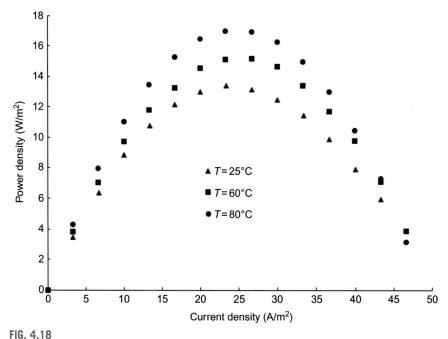

FIG. 4.18

Power density vs. current density results for the 5-cell anion exchange membrane fuel cell stack fueled with gaseous ammonia.

In the ammonia fuel cell stack tested, the voltage under no-load conditions was observed to be nearly 1.2 V when assessed at an ambient humidifier temperature. This value is marginally lower than the value expected for the 5-cell configuration considering the no-load voltage obtained for the single-cell arrangement where 278 mV were recorded. Further, the voltages under no-load conditions for humidifier temperatures of 80 and 60°C were recorded to be 1.6 and 1.4 V, respectively. Thus, increasing the temperatures of the humidifier results in no-load voltage increases of nearly 0.2 and 0.4 V, respectively. Furthermore, the power density observed at the ambient humidification temperature at the peak value was 13.4 W/m^2, which was recorded to entail a current density of 23.4 A/m^2. This power density value reflected significant losses owing to the lower power output obtained than the expected power output considering the single-cell results. However, as can be observed from the figure, the humidification temperature aided in achieving marginally high power outputs. At a higher temperature of 60°C for the humidifier, the power density at the maximum value increased by 1.8 W/m^2 to attain a peak value of 15.2 W/m^2. Also, at a higher humidification temperature of 80°C, the power output density was evaluated to be 17.0 W/m^2, which represents an increase of 3.6 W/m^2 as compared to the power output values at 25°C. The current densities under the short-circuit conditions for the 5-cell stack, however, were observed to decrease as compared to the single cell. For instance, for the single-cell arrangement, the current density was observed to be nearly 73 A/m^2 that decreases to approximately 45–50 A/m^2. This can be attributed to the high Ohmic losses that result in the stack owing to high membrane resistance. The anion exchange membrane utilized in this study entailed a high thicknesses as well as Ohmic resistance. Thus, it is recommended to investigate the performance of a DAFC stack with membranes having lower thicknesses and resistance. Also, the voltages lower than the expected values, considering the single-cell results, can be attributed to the lower electrochemical activity at the cells placed near the anodic exit. In a single-cell arrangement, the cell receives a continuous and sufficient supply of ammonia fuel. However, in a stack arrangement, the cells near the end of the stack may receive lower concentrations of fuel resulting in lower output performances from these cells. It is recommended to measure the no-load voltages across each cell in the stack during operation. This will aid in determining the effects on the fuel concentration as well as cell performances near the exit of the stack. Moreover, the energetic as well as exergetic efficiencies of the 5-cell stack evaluated at the maximum value of power output density are depicted in Fig. 4.19.

As can be observed, increasing humidifier temperatures aided in achieving both higher energetic and exergetic efficiencies. At ambient temperatures, the energy efficiency was evaluated to be 52.4% at the power density value assessed at the maximum value. This was found to increase to 55.6% when the humidifier temperature was raised to 60°C. Also, at a higher humidifier temperature of 80°C, the energy efficiency entailed a value of 66.9% evaluated at the peak value. Moreover, the exergy

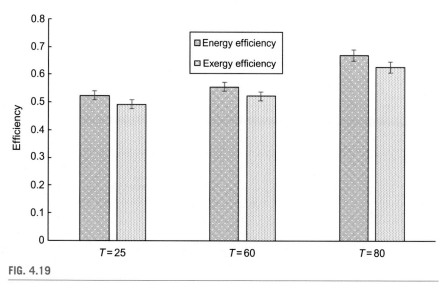

FIG. 4.19

Comparison of energy and exergy efficiencies of the 5-cell direct ammonia fuel cell stack at different humidifier temperatures.

efficiency calculated under ambient conditions was 49.3%. This is lower than the energy efficiency value found under the same conditions. Further, the exergetic efficiency at a humidification temperature of 60°C was evaluated to be 52.2%. Also, at 80°C the exergy efficiency was found to be 62.7% at the power density value assessed at the maximum points.

In addition to the anion exchange membrane DAFC and stack, the molten alkaline electrolyte cell was also investigated. A schematic representation of various components of the experimental setup is depicted in Fig. 4.20. A molten alkaline mixture of sodium and potassium hydroxide was utilized with a mole ratio of 1:1. The half-cell reactions expected in this type of cell are similar to the reactions described for an alkaline DAFC. The primary difference between the membrane-based cell and the molten electrolyte-based cell is entailed in the phase of the electrolyte. The alkaline electrolyte in a membrane-based cell is a solid-state electrolyte. However, in a molten electrolyte-based cell, the electrolyte entails a molten phase. Stainless steel tubes were used as reactant and oxidant flow inputs passages. Also, nickel coils were utilized as electrodes at both the anodic and cathodic sides. Also, they comprised an electrode area of $2.4\,cm^2$. The molten salt was contained in an alumina flask and the electrodes were immersed sufficiently.

The distance between the two electrodes was kept as low as possible. The ammonia fuel was input to the cell at a pressure of 1 bar. Also, the flow rate was kept low in this test to avoid any spillage of electrolyte. A mass flow rate of 0.2 mg/s was utilized as the anodic input where ammonia gas flows through a stainless tube to enter the

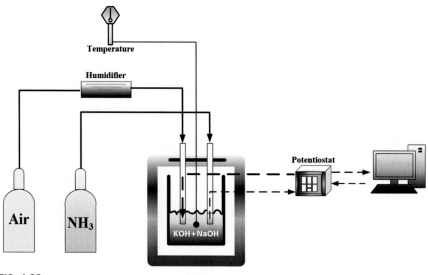

FIG. 4.20

Schematic representation of the experimental setup utilized for testing the molten alkaline electrolyte-based direct ammonia fuel cell.

electrolyte through the nickel electrode. The anodic electrochemical reaction occurs in the vicinity of the electrode where electrooxidation of ammonia occurs and electrons are emitted. Furthermore, at the cathodic side of the cell, humidified air is input to the cell to allow a continuous fuel cell operation. The molten electrolyte ammonia fuel cell was investigated at temperatures of 320, 270, and 220°C. The polarization results of the developed cell with gaseous ammonia fuel are depicted in Fig. 4.21. The voltage under no-load conditions for the molten electrolyte cell was observed to be 388 mV at the highest temperature investigated of 320°C. Also, the no-load voltage was observed to increase with decreasing operation temperatures. For instance, at a temperature of 270°C, the no-load voltage was found to drop to 484 mV and at a lower temperature of 220°C, the no-load voltage rises to 520 mV. However, the short-circuit current densities were observed to rise with increasing temperatures. Thus, as discussed earlier, further investigation on the performance of the molten electrolyte cells needs to be conducted to determine the optimal ranges of operation temperatures.

The rising current densities can be attributed to the increase in molecular as well as ionic diffusion at higher temperatures. Thus, the hydroxyl ions entail higher kinetic energies that enhances their transportability. This reduces the concentration polarization as the ions present in the vicinity of the electrodes can move with more ease. Also, at higher temperatures the Ohmic polarization is reduced owing to lower electrolyte resistances. These factors contribute to attaining higher current densities at higher operation temperatures. Moreover, the power densities assessed at varying current densities are depicted in Fig. 4.22. At operating temperatures of

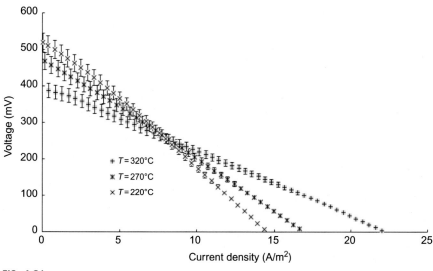

FIG. 4.21

Polarization results for the molten electrolyte DAFC.

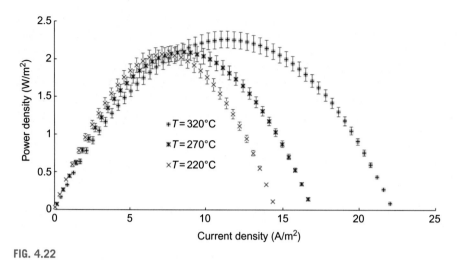

FIG. 4.22

Power density results for the molten electrolyte direct ammonia fuel cell.

320°C, the power density assessed at the peak value is observed to be nearly 2.3 W/m^2. This occurs at a current density of nearly 12.5 A/m^2. Furthermore, at lower temperatures, the maximum power densities were observed to reduce marginally. At an operating temperature of 270°C, the power density at the peak value is observed to drop to nearly 2.1 W/m^2. Also, the power density assessed at the maximum value for an operation temperature of 220°C was observed to be nearly the same value. Although the differences in the no-load voltages were observed to be significant at varying electrolyte temperatures, the power densities at the peak values were found to be similar. The current densities corresponding to this value, however, were found to be lower for lower electrolyte temperatures. In addition to this, although the no-load voltages obtained for the molten alkaline electrolyte cells are comparatively higher than the voltages obtained for the membrane electrolyte cells, the power output densities of the molten electrolyte cells are considerably lower than the power output densities of membrane utilizing cells. This is attributed to the lower short-circuit currents associated with the molten electrolyte cell. In addition to this, the power densities obtained from these cells are low and large electrode areas would be required if they are to be utilized in practical applications. However, the molten electrolyte cells can also be integrated with thermal energy storage systems that already utilize molten salts as energy storage media. In such systems, hybrid operation of the fuel cell as well as thermal storage can be implemented. This will be discussed in detail in the proceeding chapters where several integrated systems will be presented and discussed that entail the integration of DAFCs with other energy systems. In addition to this, the energy and exergy efficiencies of the molten electrolyte cell were evaluated. The variation of energy efficiency with current density is depicted in Fig. 4.23.

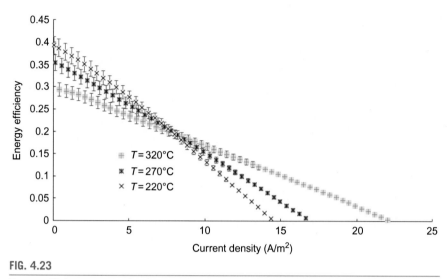

FIG. 4.23

Energy efficiency results for the molten electrolyte direct ammonia fuel cell.

In the vicinity of the power density at the maximum point, the energy efficiency is evaluated to be nearly 15% at an operation temperature of 320°C. Further, at lower temperatures, the energy efficiency at the peak power density is observed to be higher. For instance, at an electrolyte temperature of 270°C, the energetic efficiency rises to nearly 19% at the maximum point of power density. Also, at a lower temperature of 220°C, the energy efficiency increases to nearly 21%. This can be attributed to the efficiency evaluation calculation that considers the number of ammonia molecules consumed. Lower number of molecules react electrochemically at the peak power density at lower temperatures as can be observed from the lower current densities. However, although lower number of molecules are consumed, higher power output per unit molecule consumed is obtained resulting in higher energy efficiencies. Depending on the type of application and the power outputs required, the molten electrolyte ammonia fuel cell can be operated at the appropriate temperature that provides the required amount of power output as well as efficiencies. The exergy efficiency at an electrolyte temperature of 320°C is observed to be nearly 17% when assessed at the maximum value of power density. This is observed to rise to about 21% at a lower temperature of 270°C. Further, at an electrolyte temperature of 220°C the exergetic efficiency increases to around 24%. The exergy efficiencies are observed to be higher than the energy efficiencies for the molten electrolyte cell (Fig. 4.24). Also, similar trends are observed in the exergy efficiencies as were found in the energy efficiencies. Lower electrolyte temperatures provide higher efficiencies at the peak power density points. Nevertheless, the efficiencies of DAFCs developed are comparatively lower than commercially available hydrogen fuel cells. However, the usage of ammonia as fuel entails several other advantages

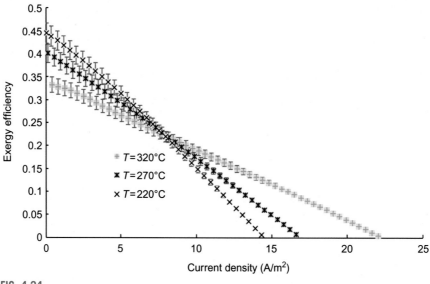

FIG. 4.24

Exergy efficiency results for the molten electrolyte direct ammonia fuel cell.

such as storage and transportation ease, lower flammability, easy leak detection, lower requirement of storage space, etc.

4.6 Ammonia borane fuel cells

In addition to different types of DAFCs, ammonia borane has also been investigated as a fuel. The anodic half-cell electrochemical reaction for these type of cells can be denoted as

$$NH_3BH_3 + 6OH^- \rightarrow BO_2^- + NH_4^+ + 4H_2O + 6e \qquad (4.19)$$

where the ammonia borane molecule (NH_3BH_3) reacts electrochemically with hydroxyl (OH^-) ions to release electrons that generate a flow of current. Also, the cathodic half-cell electrochemical reaction is similar to cathodic reaction described earlier for alkaline fuel cells where oxygen and water molecules react by accepting electrons to form hydroxyl ions. The overall reaction for direct ammonia borane fuel cells can be written as follows:

$$NH_3BH_3 + \frac{3}{2}O_2 \rightarrow BO_2^- + NH_4^+ + 4H_2O \qquad (4.20)$$

In one study [67], the electrooxidation of ammonia borane with anodic catalysts comprising palladium as well as platinum materials was studied. The catalysts developed were investigated for half-cell anodic performances. The half-cell test setup utilized the alkaline electrolyte with a molarity of 10-mM ammonia borane in 1-M sodium hydroxide. A gold plate counter electrode was employed in the 3-cell setup. This was done to prevent any hydrolysis of BH_4^- anions. Further, the reference electrode utilized comprised the reversible hydrogen electrode. The working electrodes of palladium and platinum materials were fabricated from polycrystalline metallic rotating disk electrodes. Moreover, the catalyst layer utilized to provide the active sites for the electrochemical reactions was composed of 10 wt% of platinum or palladium on carbon. Also, the carbon support is composed of Vulcan XC-72. The final MEA entailed the metallic elements on carbon cloths, a Nafion membrane, and a cathodic compartment including a platinum on carbon catalyst ($2 mg/cm^2$) on Toray carbon paper. Further, bipolar plates of graphite material were utilized that were fabricated with 25 flow channels in a serpentine configuration. The input fuel used was aqueous ammonia borane with a molarity of 1 M in 5-M sodium hydroxide solution. The oxidant utilized comprised pure humidified oxygen. The voltage under open-circuit conditions was found to be −0.052 and −0.274 V vs the reversible hydrogen electrode with platinum and palladium electrodes, respectively. Moreover, the platinum and palladium electrodes were found to entail open-circuit voltages of 0.947 and 1.050 V, respectively. Also, the power density assessed at the maximum value was observed to be 181 and $154 mW/cm^2$ for platinum and palladium electrodes, respectively. Furthermore, another study investigated aqueous ammonia borane fuel cells [68]. The developed cell entailed the usage of gold electrodes as the working electrodes for the cyclic voltammetry tests. The developed MEA was composed of 30 wt% platinum with Vulcan XC-72 at either electrodes. These were fabricated to form a low-thickness layer of nearly 100 μm that was made by mixing Nafion solution (5 wt%) with the catalyst inks. The platinum loadings were reported be approximately $0.15 mg/cm^2$ at either electrodes. Further, the assembly of electrodes and the membrane was fabricated through hot pressing of these layers coated on PTFE sheets placed on either sides of a Nafion 117 membrane. The fuel utilized for investigating the fuel cell performance was 0.01 M ammonia borane solution in 2 M sodium hydroxide solution at an input flow rate of 5 mL/min. Higher concentrations of fuel were found to provide higher open-circuit voltages. The power density was reported be higher than $14 mW/cm^2$ with platinum electrode and carbon cloth. Carbon paper was found to provide lower performances. Also, at a concentration of 0.03 M, the open-circuit voltage of −1164 mV vs SCE was observed with copper electrode. Another alkaline electrolyte entailing direct ammonia borane fuel cell was developed and tested [69]. Aqueous ammonia borane was utilized as the fuel. The MEA utilized comprised the anion exchange membrane with a thickness of 28 μm, carbon cloth on either side of the membrane, and flow channel plates made of carbon material. The platinum loading was reported to be approximately 0.93 and $0.76 mg/cm^2$ for the cathode and anode, respectively. The electrochemical characterization of the anode was performed by placing it in the fuel solution along with a mercury/mercury oxide reference electrode. To test the performance of the developed cell, aqueous ammonia borane and sodium hydroxide solution was used as fuel at a flow rate of 50 mL/min. The cell temperatures were also varied between

25 and 45°C. The oxidant supplied comprised humid oxygen at a flow rate of approximately 120 mL/min. The open-circuit potential was found to increase with rising ammonia borane concentration. Furthermore, the temperatures of the fuel solution were also varied and the performance of ammonia borane fuel cell was tested. Higher temperatures were found to enhance the cell performances marginally. Moreover, the fuel cell provided a power output higher than $110 \, mW/cm^2$ whereas the voltage and current values were recorded to be 0.6 V and $185 \, mA/cm^2$, respectively, at an operation temperature of 45°C. At lower operating temperatures, the power outputs were observed to be lower. Also, the short-circuit current densities were found to be on the lower side for fuel cell applications. To investigate the reason for high potential drops at increased current densities, the potential of both the anode and the cathode was tested at varying current densities. The cathodic potential was observed to decrease considerably as compared to the anodic potential which was fairly constant with varying current densities. Also, the study investigated the fuel crossover phenomenon which can also be identified as one of the performance lowering reasons. The cathodic catalyst layer was found to have concentrated nitrogen atoms distributed on the surface hinting a crossover of fuel. In addition to this, oxygen humidification was recommended as one of the measures that can be employed to enhance the fuel cell performance. However, the increase in cell performances was marginal.

4.7 Closing remarks

In this chapter, ammonia fuel cells are discussed comprehensively focusing on different types of ammonia fuel cells which have been investigated. The basic electrochemical equations and parameters are presented, which are needed to understand the performance of the fuel cell being discussed. The operating conditions of each type of ammonia fuel cell are discussed along with the corresponding electrodes, electrolytes, and electrochemical catalysts employed. The classification category of the type of electrolyte is deployed to categorize ammonia fuel cells according to the type of electrolytes utilized. The fundamental principles of each type of ammonia fuel cell presented are explained. Next, the performances of these ammonia fuel cells are assessed based on the open-circuit voltages, peak power densities, and the corresponding operating temperatures as well as utilized electrochemical catalysts. Furthermore, the alkaline electrolyte-based DAFCs developed at the CERL are discussed. Their development procedures are described along with the experimental methods. In addition, the fuel cell stack developed is presented along with different cell components included in the assembly. The performance of the developed fuel cells is presented in terms of the voltages, power densities, and energy and exergy efficiencies. The challenges associated with the developed fuel cells and recommendations to overcome these hindrances are also provided.

Analysis and modeling

This chapter firstly introduces fundamental thermodynamic concepts necessary to understand the thermodynamic as well as electrochemical modeling and analysis of ammonia fuel cells. The usage of these fundamental concepts for the evaluation of reversible cell potential and its underlying significance is discussed. Also, the usage of Gibb's free energy along with the Nernst equation to evaluate the reversible cell potentials is explained, and the effects of varying temperatures and pressures are presented. Next, the energy conversion efficiency of fuel cells is discussed from different aspects. Moreover, the voltage losses in fuel cells including activation, concentration, and Ohmic losses are covered comprehensively. The underlying phenomena that lead to these losses are also described in detail.

5.1 Thermodynamics of ammonia fuel cells

In this chapter, firstly the thermodynamic concepts of enthalpy are covered. This includes absolute enthalpy and its variation with changing operation parameters such as temperatures and pressures. Furthermore, enthalpy of formation, as well as combustion, is discussed. In addition to this, the concepts of lower, heating value and higher heating value (HHV) are covered. Next, the Gibbs function of a reaction is described for ammonia fuel cells that is an essential parameter to be calculated while evaluating the reversible or Nernst cell potentials.

In systems including chemical reactions, the evaluation of absolute enthalpies is essential for thermodynamic analysis and assessment. The absolute enthalpy entails the combination of sensible thermal energy as well as chemical energy. The sensible thermal energy denotes the enthalpy of substance with reference to a given state. The chemical energy comprises the energy entailed in the chemical bonds of a given chemical specie. This is also often denoted by the enthalpy of formation. Enthalpy is a function of the temperature as well as the pressure. As these two intensive properties vary, the enthalpies also change. The absolute enthalpy for a given substance can be defined as

$$h(P,T) = h_f\left(P_{ref}, T_{ref}\right) + \Delta h_{sen}(P,T) \qquad (5.1)$$

where h denotes the absolute enthalpy, h_f denotes the enthalpy of formation, and Δh_{sen} represents sensible thermal energy. Also, P and T denote the pressure and

Ammonia Fuel Cells. https://doi.org/10.1016/B978-0-12-822825-8.00005-0

123

temperature, respectively. The enthalpy of formation can be found either through experimental techniques or by thermochemical analysis. In addition, the reference pressure and temperature can be chosen according to the reference conditions. However, owing to tabulated thermodynamic parameters, the reference pressure and temperature of 1 atm and 25°C are generally utilized. Moreover, in thermodynamics of chemical reactions, the elemental enthalpy is considered to be zero in their natural states at the reference conditions. For instance, the enthalpy of formation of hydrogen, oxygen, nitrogen, etc. is zero when they are at a temperature of 25°C and pressure of 1 atm. However, other molecules entailing more than one element entail enthalpies of formation. Carbon dioxide, for example, entails an enthalpy of formation of −393,522 J/mol that represents the amount of heat emitted when 1 mol carbon dioxide is formed after reacting with 1 mol of carbon and 1 mol oxygen. The negative symbol with the enthalpy denotes the emission of heat rather than absorption of heat, which entails a positive symbol with the enthalpy of formation. Currently, the enthalpy of formation at reference conditions for various substances is tabulated in various thermodynamic tables.

The enthalpy of reaction in a given chemical system can be identified as the difference between the total enthalpy of products and the total enthalpy of reactants. To determine the enthalpy of reaction experimentally, the thermal energy released during the reaction while operating isobarically can be recorded. Such experiments can be conducted under closed as well as open systems. In closed systems, the pressures and temperatures at the initial state before the reaction must equal the pressures and temperatures at the final state after the reaction. In open systems, the temperatures and pressures of the inlet flows must equal to the temperatures and pressures of the outlet flow. In either system, if heat is absorbed by the control volume, the amount of heat absorbed can be evaluated as

$$q = h_{prod} - h_{reac} \tag{5.2}$$

where the heat absorbed (q) also denotes the enthalpy of reaction. However, some chemical reactions entail the release of heat while some include the absorbing of heat. In case of heat release, the reaction is classified as exothermic and heat-absorbing reactions are denoted as endothermic reactions. Thus, in endothermic reactions, an external heat input source is required whereas in exothermic reactions heat is released from the reaction that is generally required to be taken out of the system to avoid the operating temperature from increasing significantly. When a fuel reacts with an oxidant in an exothermic manner, the enthalpy of reaction is also denoted as the enthalpy of combustion. As the enthalpy of the reactant or fuel as well the oxidant is dependent on the pressure as well as temperature, the enthalpy of combustion is also dependent on these factors. For instance, when methane (CH_4) is combusted at standard conditions (1 atm and 25°C), carbon dioxide and water are produced as a result of complete combustion. The water molecules produced can be in either the liquid state or gaseous state in their final states. If the water molecules condense and attain the liquid state, the enthalpy of combustion is referred to as the

HHV. When the water molecules do not condense and remain in the gaseous state, the enthalpy of combustion is referred to as the lower heating value (LHV). Hence, the difference between these two heating values entails the enthalpy of vaporization of water. For instance, the combustion reaction of ammonia to form nitrogen and water can be written as

$$NH_3 + \frac{3}{4}O_2 \rightarrow \frac{1}{2}N_2 + \frac{3}{2}H_2O \tag{5.3}$$

The enthalpy of reaction for the ammonia combustion reaction showed above can be evaluated as

$$\Delta h_{reac} = \frac{1}{2}h_{N_2} + \frac{3}{2}h_{H_2O} - \left(h_{NH_3} + \frac{3}{4}h_{O_2}\right) \tag{5.4}$$

Through these equations, the HHV and LHV of ammonia can be determined as 317.56 and 382.6 kJ/mol, respectively, when evaluated at standard conditions.

In addition to enthalpy, another thermodynamic property that is utilized extensively in the analysis of electrochemical as well as chemical systems is the Gibbs function that is defined as follows:

$$g = h - Ts \tag{5.5}$$

where g denotes the Gibbs function, h represents the specific enthalpy, s denotes the specific entropy, and T is the temperature. As previously, the enthalpy of reaction was evaluated from the enthalpies of reactants and products, similarly the change in Gibbs function of reaction is also calculated as the difference between the Gibbs function of the products and reactants when they both entail the same pressure as well as temperature.

$$\Delta g_{reac} = g_{prod} - g_{reac} \tag{5.6}$$

Similar to the enthalpy of formation, when the change in Gibbs function is evaluated for the formation of a compound, it is referred to as the Gibbs function of formation.

During a fuel cell operation, the chemical energy entailed in the fuel as well as oxidant converts into electrical energy directly. This is associated with an electrical potential as well as current output. If the operation is assumed to be conducted under thermodynamically reversible conditions, the maximum amount of useful output energy possible can be evaluated. Similarly, the potential of a cell under reversible conditions is referred to as the reversible cell potential that also corresponds to the maximum possible potential that can be attained. This is an essential parameter in the analysis of fuel cell systems. Consider an operating fuel cell as a control volume system where the fuel enters the anodic compartment and the oxidant enters the cathodic side. Also, the products leave from the exit stream from the cell. The cell is assumed to be placed in an isothermal bath that maintains its temperature at a given value. Further, the temperatures and pressures of all inlet and exit streams are assumed to be identical. Also, the changes in the potential, as well as kinetic energies, are taken to be negligibly small. The energy balance can be applied to such a fuel cell system as follows:

$$\dot{N}_f \overline{h}_f + \dot{N}_{ox} \overline{h}_{ox} + \dot{Q} = \frac{dE_{cv}}{dt} + \dot{N}_{ex} \overline{h}_{ex} + \dot{W} \qquad (5.7)$$

where $\dot{N}_f$ denotes the molar flow rate of input fuel, $\overline{h}_f$ represents the specific molar enthalpy of the input fuel, $\dot{N}_{ox}$ is the molar flow rate of oxidant, $\overline{h}_{ox}$ is the specific molar enthalpy of the oxidant, $\dot{Q}$ is the heat input to the control volume, $\frac{dE_{cv}}{dt}$ denotes the change in the energy of control volume with time, $\dot{N}_{ex}$ is the molar flow rate of the exit stream, $\overline{h}_{ex}$ denotes the specific molar enthalpy of the exhaust stream, and $\dot{W}$ is the power output from the cell.

Similarly, the entropy balance can also be applied to the control volume as

$$\dot{N}_f \overline{s}_f + \dot{N}_{ox} \overline{s}_{ox} + \frac{\dot{Q}}{T} + \dot{S}_{gen} = \frac{dS_{cv}}{dt} + \dot{N}_{ex} \overline{s}_{ex} \qquad (5.8)$$

where $\overline{s}_f$ denotes the specific molar entropy of the fuel, $\overline{s}_{ox}$ represents the specific molar entropy of the oxidant, $\dot{S}_{gen}$ is the rate of entropy generation within the control volume, $\frac{dS_{cv}}{dt}$ denotes the change in entropy of the control volume with time, and $\overline{s}_{ex}$ is the specific molar entropy of the exhaust stream. However, under steady-state conditions there is no change in the amount of energy or entropy of the system. Thus, the energy and entropy balance for such cases can be written, respectively, as follows:

$$\dot{N}_f \overline{h}_f + \dot{N}_{ox} \overline{h}_{ox} + \dot{Q} = \dot{N}_{ex} \overline{h}_{ex} + \dot{W} \qquad (5.9)$$

$$\dot{N}_f \overline{s}_f + \dot{N}_{ox} \overline{s}_{ox} + \frac{\dot{Q}}{T} + \dot{S}_{gen} = \dot{N}_{ex} \overline{s}_{ex} \qquad (5.10)$$

In addition to this, the enthalpies at the inlet and exit in the above equations can be written per unit mole of fuel to obtain

$$\overline{h}_{in} = \overline{h}_f + \frac{\dot{N}_{ox}}{\dot{N}_f} \overline{h}_{ox} \qquad (5.11)$$

$$\overline{h}_{out} = \frac{\dot{N}_{ex}}{\dot{N}_f} \overline{h}_{ex} \qquad (5.12)$$

Also, the entropies at the inlet and exit can be written per unit mole of fuel as follows:

$$\overline{s}_{in} = \overline{s}_f + \frac{\dot{N}_{ox}}{\dot{N}_f} \overline{s}_{ox} \qquad (5.13)$$

$$\overline{s}_{out} = \frac{\dot{N}_{ex}}{\dot{N}_f} \overline{s}_{ex} \qquad (5.14)$$

With these enthalpies and entropies, the energy and entropy balance equations can be rewritten as

$$\dot{N}_f\left(\overline{h}_{in}-\overline{h}_{out}\right)+\dot{Q}=\dot{W} \tag{5.15}$$

$$\dot{N}_fT\left(\overline{s}_{in}-\overline{s}_{out}\right)+T\dot{S}_{gen}=-\dot{Q} \tag{5.16}$$

Furthermore, substituting Eq. (5.16) in Eq. (5.15) gives

$$\dot{N}_f\left(\overline{h}_{in}-\overline{h}_{out}\right)-\left(\dot{N}_fT\left(\overline{s}_{in}-\overline{s}_{out}\right)+T\dot{S}_{gen}\right)=\dot{W} \tag{5.17}$$

Moreover, the entropy generation, work output as well as heat input terms can be written per unit mole of fuel as

$$\frac{\dot{W}}{\dot{N}_f}=w \tag{5.18}$$

$$\frac{\dot{Q}}{\dot{N}_f}=q \tag{5.19}$$

$$\frac{\dot{S}_{gen}}{\dot{N}_f}=s_{gen} \tag{5.20}$$

The energy and entropy balance equations can be rewritten by substituting the above equations as

$$\overline{h}_{in}-\overline{h}_{out}+\left(T\left(\overline{s}_{in}-\overline{s}_{out}\right)-Ts_{gen}\right)=w \tag{5.21}$$

$$-T\left(\overline{s}_{in}-\overline{s}_{out}\right)-Ts_{gen}=T\Delta s-Ts_{gen}=q \tag{5.22}$$

Further, Eq. (5.15) can be rewritten in the following form:

$$-\Delta h+T\,\Delta s-Ts_{gen}=w \tag{5.23}$$

where Δh denotes the difference between the enthalpy at the exit $(\overline{h}_{out})$ and the enthalpy at the inlet $(\overline{h}_{in})$. Similarly, Δs denotes the entropy difference between the exit $(\overline{s}_{out})$ and the inlet $(\overline{s}_{in})$. The above equation can also be rewritten as follows:

$$-\left\{(h-Ts)_{out}-(h-Ts)_{in}\right\}-Ts_{gen}=w \tag{5.24}$$

Next, Eq. (5.5) can be substituted in the above equation to obtain an expression for the specific work in terms of the Gibbs function as

$$-\left\{(g)_{out}-(g)_{in}\right\}-Ts_{gen}=-\Delta g-Ts_{gen}=w \tag{5.25}$$

In real systems, the entropy generation is always positive according to the second law of thermodynamics. Entropy can be generated but cannot be destroyed. However, in ideal systems, the entropy generated is assumed to be zero. Such systems are referred to as reversible systems and do not entail any irreversibilities or generation of entropy. Hence, if the system that is being considered in the equations above is taken to be a reversible system, the equation for work output can be written as

$$-\Delta g = w_{rev} \qquad (5.26)$$

where the reversible work denoted by w_{rev} also corresponds to the maximum possible work that can occur when the system operates with no irreversibilities. Therefore, for a given fuel cell system the maximum possible work can be denoted from the above equation. Moreover, one of the important parameters in fuel cell systems is the potential difference between the anode and cathode that is generally expressed in terms of volts (V). The potential difference refers to the electrical potential energy per coulomb of electrical charge (J/C). The electrical potential energy refers to the work done in moving a charge between two given points in an electrical field. This is generally used for external circuits. However, in cases such as fuel cells or batteries, internal circuits exist and the term electromotive force is generally used. Nevertheless, the term "cell potential" is generally used for fuel cells. In fuel cells, electrons are emitted at the anode and accepted at the cathode that generates a flow of charge across an external circuit. The work output of a fuel cell can thus be expressed as

$$w = (E)(C) \qquad (5.27)$$

where w denotes the work output per unit mole of fuel, E denotes the potential of the cell and C is the charge. Writing the charge (C) in terms of the Faraday's constant (F) and the number of moles of transferred electrons (n), the above equation can be written as follows:

$$w = (E)(nF) \qquad (5.28)$$

Replacing the work output term with Eq. (5.19), the above equation can be rearranged to solve for

$$E = \frac{-\Delta g - T s_{gen}}{nF} \qquad (5.29)$$

In the case of a reversible system where no entropy generation occurs, the cell potential can be evaluated as

$$E_{rev} = \frac{-\Delta g}{nF} \qquad (5.30)$$

where E_{rev} denotes the reversible cell potential that is obtained when a fuel cell operates with no entropy generation or irreversibilities. Further, the amount of voltage loss due to entropy generation can be subtracted from the reversible cell potential:

$$E = E_{rev} - \frac{T s_{gen}}{nF} \qquad (5.31)$$

Moreover, rewriting Eq. (5.6) the Gibbs function can also be written as

$$\Delta g = \Delta h - T \Delta s \qquad (5.32)$$

As can be observed from the equation, the Gibbs function is dependent on properties such as temperature, pressure, etc. that directly effect the enthalpies and entropies of a given state. The reversible cell potential of direct ammonia fuel cells at reference conditions such as 25°C and 100 kPa, can be evaluated from the above equations as 1.17 V.

Table 5.1 Comparison of different parameters in the evaluation of the reversible cell potential of ammonia and hydrogen fuel cells at reference conditions.

Fuel utilized	Reversible cell potential (E_{rev})	Change in Gibbs function ($-\Delta g$)	Change in molar enthalpies ($-\Delta \bar{h}$)	Moles of electrons (n)	Overall fuel cell reaction
Ammonia	1.17 V	338.2 J/mol	382.8 J/mol	3	$NH_3 + \frac{3}{4}O_2 \rightarrow \frac{3}{2}H_2O + \frac{1}{2}N_2$
Hydrogen	1.229 V	237.3 J/mol	286.0 J/mol	2	$H_2 + \frac{1}{2}O_2 \rightarrow H_2O$

Also, the reversible cell potential of hydrogen fuel cells at these reference conditions can be evaluated as 1.229 V. Furthermore, Table 5.1 summarizes the evaluated parameters as well as the overall reactions for these fuel cells. The reversible cell potential, as well as other parameters listed, is evaluated at the reference conditions.

In fuel cells, the primary operation parameters that effect the performance of the cell are pressure, temperature, and the concentration of species. Prior to discussing how this main parameter would change the cell performance, few thermodynamic relations are formulated. As discussed earlier, the enthalpy and Gibbs function are denoted as

$$h = u + Pv \tag{5.33}$$

$$g = h - Ts \tag{5.34}$$

Substituting Eq. (5.26) into Eq. (5.27) to obtain the following:

$$dg = du + Pdv + vdP - Tds - sdT \tag{5.35}$$

Furthermore, for a given compressible substance the following thermodynamic relation can be applied:

$$Tds = dh - vdP = du + Pdv \tag{5.36}$$

Next, if we substitute the above equation into Eq. (5.28) we obtain

$$dg = vdP - sdT \tag{5.37}$$

Thus, as far as fuel cells are concerned, two thermodynamic relations can be derived from the above equation:

$$-s = \left(\frac{\partial g}{\partial T}\right)_P \tag{5.38}$$

$$v = \left(\frac{\partial g}{\partial P}\right)_T \tag{5.39}$$

Moreover, we can implement the change in the Gibbs function, change in specific volume and entropy to rewrite the above equations as follows:

$$-\Delta s = \left(\frac{\partial \Delta g}{\partial T}\right)_P \tag{5.40}$$

$$\Delta v = \left(\frac{\partial \Delta g}{\partial P}\right)_T \tag{5.41}$$

As can be observed from the above equations, temperatures and pressures can be varied to analyze their effects on the fuel cell performances. Recall that the reversible cell potential can be expressed in terms of the change in Gibbs function and both are dependent on the temperatures and pressures:

$$E_{rev}(P,T) = -\frac{\Delta g(P,T)}{nF} \tag{5.42}$$

Taking the partial differentials of both sides with respect to temperature, the above equation can be expressed as

$$\left(\frac{\partial E_{rev}(P,T)}{\partial T}\right)_P = -\frac{1}{nF}\left(\frac{\partial \Delta g(P,T)}{\partial T}\right)_P \tag{5.43}$$

Furthermore, Eq. (5.33) can be substituted in the above equation to obtain

$$\left(\frac{\partial E_{rev}(P,T)}{\partial T}\right)_P = \frac{\Delta s(P,T)}{nF} \tag{5.44}$$

Thus, as can be observed the change in the reversible cell potential with temperature is dependent on the change in the entropy for a given fuel cell. Hence, for a given type of fuel cell, the reversible potential will drop with temperature if $\Delta s < 0$. Also, it will increase with a rise in temperature if $\Delta s > 0$ and it does not change with temperature if $\Delta s = 0$. Furthermore, Eq. (5.37) can be integrated with respect to temperature with limits of a reference temperature and operating temperature considering the change in entropy is nearly constant with change in temperature as in many electrochemical reactions to obtain:

$$E_{rev}(P,T) = E_{rev}(P,T_{ref}) + \left(\frac{\Delta s(P,T_{ref})}{nF}\right)(T-T_{ref}) \tag{5.45}$$

However, the above equation depicts an approximation of the reversible cell potential and Eq. (5.23) should be utilized to evaluate the reversible cell potential. Also, the effect of pressure on the fuel cell potential can be analyzed through Eqs. (5.34) and (5.36) to obtain:

$$\left(\frac{\partial E_{rev}(P,T)}{\partial P}\right)_T = -\frac{1}{nF}\left(\frac{\partial \Delta g(P,T)}{\partial P}\right)_T \tag{5.46}$$

where the term on the right-hand side can be rewritten by substituting Eq. (5.34) to obtain

$$\left(\frac{\partial E_{rev}(P,T)}{\partial P}\right)_T = -\frac{\Delta v}{nF} \tag{5.47}$$

where the change in the specific volume denotes the difference between the specific volumes of product (v_{prod}) and the reactants (v_{reac}):

$$v_{prod} - v_{reac} = \Delta v \tag{5.48}$$

Treating the products and reactants as ideal gases, the above equation can also be written as

$$\Delta v = \frac{\Delta NRT}{P} \tag{5.49}$$

where ΔN can be expressed as follows:

$$\Delta N = N_{prod} - N_{reac} \tag{5.50}$$

Further, these equations can be used to rewrite Eq. (5.40) as

$$\left(\frac{\partial E_{rev}(P,T)}{\partial P}\right)_T = -\frac{1}{P}\frac{\Delta NRT}{nF} \tag{5.51}$$

Hence, as can be observed from the above equation, when the difference between the number of moles of product molecules is higher than the number of moles of reactants ($\Delta N > 0$), the reversible potential is expected to drop with the operating pressure. Furthermore, if the moles of reactant gas molecules are higher than the moles of product molecules ($\Delta N < 0$), the reversible cell potential entails an increasing trend with pressure. Also, in cases where the equal number of gas molecules exists on both the reactant as well as the product side, the reversible potential is independent of pressure change. In addition to this, operating at high pressures entails other types of challenges such as mechanical issues, limited strength of cell components, more gas leakage, etc.

Integration of Eq. (5.44) between the limits of the reference pressure and the operation pressure considering constant temperature leads to the following:

$$E_{rev}(P,T) = E_{rev}(P_{ref},T) - \frac{\Delta NRT}{nF}\ln\left(\frac{P}{P_{ref}}\right) \tag{5.52}$$

As can be observed from the above equation, at a constant temperature the reversible potential varies with pressure in a logarithmic manner. Thus, as the operation pressure (P) increases, the amount of effect on the reversible cell potential becomes feebler.

Along with operating temperatures and pressure, the concentrations also effect the reversible cell potentials. The equations derived so far were based on the assumption that both the fuel as well as the oxidant streams are pure along with a pure exhaust stream. However, other substances are generally entailed in these streams rather than a pure feed or exhaust. For instance, the ammonia fuel input to the cell can be humidified to provide moisture to the membranes. The fuel stream thus comprise both ammonia as well as water molecules at different concentrations depending on the amount of humidification applied. Furthermore, in case of hydrogen fuel cells, if the fuel is derived from steam methane, reforming it comprises other chemical species including carbon dioxide, carbon monoxide, water, etc. In addition to this, in fuel cell applications the oxidant is usually supplied in the form of air. The oxygen contained in the air is utilized to operate the fuel cell. Hence, the oxidant input stream comprises other chemical compounds such as nitrogen, water vapor, etc. This leads to a particular oxygen concentration at the fuel cell inlet and the stream does not entail pure oxygen. In such cases where the fuel and oxidant streams are not completely pure and entail other compounds, their partial pressures in the overall mixture can be utilized in the calculation of the Gibbs functions. Thus, the Gibbs function for a given component "a" can be denoted for constant temperature as

$$g_a(P_a, T) = h_a(T) - T s_a(P_a, T) \tag{5.53}$$

Furthermore, Eq. (5.29) can be rewritten considering a specific specie as

$$ds_a = \frac{dh_a}{T} - \frac{v}{T} dP \tag{5.54}$$

The above equation can be further rewritten by applying the ideal gas law as

$$ds_a = \frac{dh_a}{T} - \frac{R}{P} dP \tag{5.55}$$

Next, integrating Eq. (5.48) between the limits of the reference pressure (P_{ref}) and the operating pressure (P) considering isothermal operation gives

$$s_a(P, T) - s_a(P_a, T) = -R \left(\ln P - \ln P_a \right) \tag{5.56}$$

This can also be written by combining the logarithmic terms as

$$s_a(P, T) - R \ln \left(\frac{P_a}{P} \right) = s_a(P_a, T) \tag{5.57}$$

Furthermore, the Gibbs function denoted in Eq. (5.46) can be rewritten by substituting Eq. (5.50) as

$$g_a(P_a, T) = h_a(T) - T s_a(P, T) + RT \ln \left(\frac{P_a}{P} \right) \tag{5.58}$$

where the g_a denotes the Gibbs function of specie (a) in case of a mixture, T represents the temperature of the mixture, and P denotes the mixture pressure.

However, the reversible cell potential denotes the maximum potential that occurs only under reversible conditions. In real fuel cells, there are several types of losses and irreversibilities that lead to voltage losses. In addition, fuel cells are generally made to operate near their maximum point of power output densities. These occur in the vicinity of particular values of voltage and current. As current is drawn from a fuel cell, the amount of irreversibilities increase that are associated with different types of voltage losses. These are referred to as polarization losses that are discussed comprehensively in the proceeding chapters.

5.2 Electrochemical analysis of ammonia fuel cells

The previous chapter entailed a discussion about the thermodynamic analysis of fuel cells and the evaluation of the reversible cell potential that occurs under reversible operation. However, in order to determine the rate of an electrochemical reaction further analysis of electrochemical processes occurring within the cell is required. In addition to this, electrochemical analysis allows us to determine how the reactants are interacting to form the products through the pathway of a given half-cell reaction. Moreover, it is essential to determine the amount of voltage losses that occur as the current is drawn from the cell to evaluate the cell voltage at a given amount of current output. This is also essential to calculate the power output from the fuel cell that corresponds to the actual useful output from the fuel cell. Also, determining different types of losses and irreversibilities aids in identifying the location of their occurrence and ways to overcome these in a given fuel cell. The rate at which an electrochemical reaction occurs determines the rate at which useful energy can be extracted from the fuel cell. In an electrochemical analysis, both the anode as well as the cathode are analyzed as both half-cell interactions effect the fuel cell performance.

As discussed earlier, in alkaline direct ammonia fuel cells the anodic reaction can be expressed as

$$NH_3 + 3OH^- \rightarrow \frac{1}{2}N_2 + 3H_2O + 3e \qquad (5.59)$$

Also, the cathodic half-cell reaction can be written as

$$\frac{3}{4}O_2 + 1.5H_2O + 3e \rightarrow 3OH^- \qquad (5.60)$$

In addition, the overall reaction is expressed as follows:

$$NH_3 + \frac{3}{4}O_2 = \frac{1}{2}N_2 + 1.5H_2O \qquad (5.61)$$

In the previous discussion about the cell potential of an operating fuel cell, the fuel cell potential was referred to as the difference between the cathodic and anodic potentials. However, each electrode entails a half-cell potential that can be measured in comparison with a given reference electrode. This is essential to study the electrochemical kinetics at the anode or cathode of a given fuel cell. Considering the reversible potential as discussed earlier (E_{rev}), this can be denoted in terms of the cathodic and anodic reversible half-cell potentials as

$$E_{rev} = \epsilon_{c,r} - \epsilon_{a,r} \qquad (5.62)$$

where $\epsilon_{c,r}$ denotes the half-cell potential of the cathode under reversible conditions and $\epsilon_{a,r}$ represents the reversible half-cell potential at the anode. Similarly, the actual cell potential can be expressed as follows:

$$E = \epsilon_c - \epsilon_a \qquad (5.63)$$

Utilizing these two equations the total overpotential occurring in a given fuel cell can be evaluated as

$$\eta_{op} = E_{rev} - E \tag{5.64}$$

Eqs. (5.55) and (5.56) can be substituted in the above equation to obtain the following relation:

$$\eta_{op} = (\epsilon_a - \epsilon_{a,r}) - (\epsilon_c - \epsilon_{c,r}) = \eta_{op,a} - \eta_{op,c} \tag{5.65}$$

where $\eta_{op,a}$ denotes the total anodic overpotential and $\eta_{op,c}$ represents the total cathodic overpotential. In fuel cell applications, the overpotential at the anode generally entails a positive value whereas the overpotential at the cathode is associated with a negative value. These result due to the convention in electrochemical analyses where the overpotential at a given electrode is measured between the actual and the reversible potentials. Thus, the potential of a given fuel cell can be determined in terms of the reversible cell potential and the overpotentials as

$$E = E_{rev} - \left(\eta_{op,a} - \eta_{op,c}\right) \tag{5.66}$$

The above equation can also be rewritten as follows:

$$E = E_{rev} - \sum |\eta_{op,i}| \tag{5.67}$$

where $\eta_{op,i}$ denotes the overpotential and i represents the anode (a) or cathode (c), respectively. For direct ammonia alkaline fuel cells entailing the half-cell reactions discussed earlier, the half-cell anodic and cathodic potentials are listed in Table 5.2. Also, to provide a comparison of the half-cell potentials alkaline hydrogen fuel cells are also tabulated.

In addition to this, the reversible cell potential of a fuel cell described earlier denotes the maximum possible cell potential that can be attained under reversible conditions. Also, the maximum cell potential is achieved when no current is drawn from the cell. This is also referred to as the open-circuit potential (E_{OCP}). The open-circuit voltage that is measured experimentally for an actual fuel cell is lower than the reversible cell potential. This difference is usually marginal and arises due to the presence of irreversibilities.

The half-cell electrochemical reactions presented earlier are referred to as the overall or global half-cell reactions. However, in actual operation, these are not single-step

Table 5.2 Half-cell electrochemical reactions and corresponding potentials of alkaline ammonia and hydrogen fuel cells.

Electrochemical reaction	Half-cell potential at standard conditions (V)
$2NH_3 + 6OH^- \rightarrow N_2 + 6H_2O + 6e$	−0.77
$O_2 + 2H_2O + 4e \rightarrow 4OH^-$	+0.40
$H_2 + 2OH^- \rightarrow 2H_2O + 2e$	−0.828

reactions where the ammonia, hydrogen, or oxygen molecules interact directly with ions and electrons. Instead, several reaction steps are involved that entails a series of interactions at either electrode. These reactions are commonly known as elementary reactions. For the electrooxidation of ammonia on the anode, the following elementary reactions have been proposed for direct ammonia alkaline fuel cells:

$$NH_3 + * \leftrightarrow NH_{3(AD)} \tag{5.68}$$

$$NH_{3(AD)} + OH^- \leftrightarrow NH_{2(AD)} + H_2O + e \tag{5.69}$$

$$NH_{2(AD)} + OH^- \leftrightarrow NH_{(AD)} + H_2O + e \tag{5.70}$$

$$NH_{(AD)} + OH^- \leftrightarrow N_{(AD)} + H_2O + e \tag{5.71}$$

where the ammonia molecule (NH_3) is first adsorbed followed by consequential removal of hydrogen atoms from the molecule that reacts with the hydroxyl (OH^-) anions. This interaction generates electrons as well as water molecules. Thus, the combination of these elementary steps leads to the global overall half-cell anodic electrochemical reaction. In addition to this, although these elementary steps have been proposed for the electrooxidation of ammonia, there can be various types of elementary reactions that occur for a given overall global electrochemical reaction.

During a fuel cell operation, the electrochemical reactions take place at the interface of the electrolyte and the electrode. The reaction rate in cases of a heterogeneous reaction is generally analyzed as moles per unit surface area of electrode as well as time. For a given elementary reaction, the general balance can be applied as

$$\sum_{n=1}^{N} \gamma'_n M_n \rightarrow \sum_{n=1}^{N} \gamma''_n M_n \tag{5.72}$$

Utilizing the above equation, the rate of formation of a given specie (n) is denoted by the mass action law as follows:

$$\tau''_{n,f} = (\gamma''_n - \gamma'_n) K_f \prod_{n=1}^{N} [M_n]^{\gamma'} \tag{5.73}$$

where M_n denotes the molar concentration that can also be evaluated using the ideal gas equation as

$$M_n = \frac{P_n}{RT} \tag{5.74}$$

Here, the exponent on the molar concentration term in Eq. (5.66) denotes the order of the reaction for a given specie. Also, the total order of the reaction can be evaluated as

$$\gamma = \sum_{n=1}^{N} \gamma'_n \tag{5.75}$$

Moreover, the reaction rate constant is denoted by K that is dependent on various factors including operating temperature, type, and material of electrolyte, type of catalyst as well as active surface, etc. Furthermore, the rate of reaction denoted by Eq. (5.66) can also be written in terms of the fuel cell current, which is easier to obtain through experiments. Generally, the current across the fuel cell can be denoted by Faraday's law. Utilizing this relation, the reaction rate and the current can be related as

$$-nF\left(\tau_F''\right) = -\left(\gamma_F'' - \gamma_F'\right)K\prod_{n=1}^{N}[M_n]^{\gamma'} = J \tag{5.76}$$

where F represents the Faraday's constant and n denotes the number of moles of electrons transferred. As can be observed, the relation depicted by Eq. (5.70) entails a negative sign that relates to the decrease in the number of moles of fuel with an increase in the current. In addition, the elementary reactions are also associated with backward reaction along with the forward reaction. The backward reaction also entails a reaction rate constant and the rate of this backward reaction can be expressed in a similar way as the forward reaction:

$$\tau_{n,b}'' = \left(\gamma_n' - \gamma_n''\right)K_b\prod_{n=1}^{N}[M_n]^{\gamma''} \tag{5.77}$$

Incorporating these forward and backward rates of reaction, the net rate of reaction can be expressed for a given specie n as follows:

$$\tau_n'' = \tau_{n,f}'' + \tau_{n,b}'' \tag{5.78}$$

The above equation can be further expanded as

$$\tau_n'' = \left(\gamma_n'' - \gamma_n'\right)\left[K_f\prod_{n=1}^{N}[M_n]^{\gamma'} - K_b\prod_{n=1}^{N}[M_n]^{\gamma''}\right] \tag{5.79}$$

When a reaction proceeds and attains equilibrium, both the rate of backward as well as forward reactions attain the same value. In such cases, the net rate of reaction becomes insignificant and the above equation can be rewritten as

$$\frac{K_f}{K_b} = \prod_{n=1}^{N}[M_n]^{(\gamma'' - \gamma')} \tag{5.80}$$

where the ratio of the forward and backward reaction rate constants can be termed as the concentration equilibrium constant. It is a function of both the temperature as well as the pressure:

$$\frac{K_f}{K_b} = K_c(P,T) \tag{5.81}$$

where K_c denotes the concentration equilibrium constant. Also, the partial pressure equilibrium constant can be written as

$$K_P = \prod_{n=1}^{N}\left[\frac{P_n}{P_0}\right]^{(\gamma'' - \gamma')} \tag{5.82}$$

The above equation can also be rewritten as

$$K_P = \left(\frac{RT}{P_0}\right)^{\sum\limits_{n=1}^{N}(\gamma''-\gamma')} \prod_{n=1}^{N} [C_n]^{(\gamma''-\gamma')} \tag{5.83}$$

As can be observed from the above equation, the partial pressure equilibrium constant depends on the temperature. This can also be found from the Gibbs reaction function as

$$K_P(T) = \exp\left(-\frac{\Delta g(P_0, T)}{RT}\right) \tag{5.84}$$

Moreover, in cases where the reactant moles equal the number of product moles in gaseous phase, the partial pressure equilibrium constant can be equated to the concentration equilibrium constant:

$$K_P = K_c \tag{5.85}$$

As can be observed from the above equation, the concentration equilibrium constant will entail a dependency solely on the temperature in such cases.

In addition to this, as a given half-cell electrode electrochemical reaction takes place at the surface of the electrode, it is essential to analyze the coverage of the electrode surface by the interacting species. The chemical species interacting in the electrochemical reactions firstly are adsorbed on the electrode surface. This makes the surface of the electrode layered with adsorbed species. However, the electrode surface may not be completely covered only with the adsorbed species of the reactants. There can be several other types of chemical species that may be adsorbed on the surface. In addition, the adsorption process may not completely occur and partial processes also occur in parallel. The reaction rate of the electrochemical interaction at the electrode is effected by these factors. Also, the presence of vacant reaction sites on the surface of the electrode influence the reaction rate. Thus, the significance of these subprocesses of desorption as well adsorption is denoted through the amount of the surface of the electrode covered with adsorbed species. This is also generally referred to as the surface coverage. An expression for the surface coverage can be written as follows:

$$\mu_n = \frac{C_{n,ads}}{\sum\limits_{k}(C_{k,ads})_s} \tag{5.86}$$

where μ_n denotes the surface coverage, $C_{n,ads}$ represents the adsorbed specie concentration and the denominator denotes the electrode saturation concentration. However, it does not always denote the coverage of the surface completely. Moreover, for the purpose of denoting the reaction rate in a correct manner, it is essential to express the electrochemical reactions as denoted in Eqs. (5.61)–(5.64). In the case of ammonia fuel cells, Eq. (5.79) can be used to evaluate the concentration of adsorbed ammonia species ($C_{NH_3,ADS}$). Utilizing these, the reaction rate equation can also be denoted as follows:

$$\tau''_{NH_3,ADS} = -k_{2,f}\mu_{NH} + k_{2,b}C_{NH} + C_e - (1 - \mu_{NH}) \tag{5.87}$$

In the above equation, the term $(1 - \mu_{\mathrm{NH}})$ denotes the surface of the electrode remaining uncovered from the layering of adsorbed ammonia. In fuel cell applications, heterogeneous reactions occur whereas other homogenous chemical reactions also exist that involve only one phase. The rate of reaction equations varies according to the type of reactions.

As discussed earlier, there are several other types of elementary reactions that can occur in parallel in a given type of fuel cell. Thus, any given overall reaction can be analyzed from its rate-determining step of reaction, which is defined as the step in the series of reactions that makes the overall reaction rate rely on the rate of this rate-determining step. Also, this is generally considered to be the case for parallel as well as consecutive reactions including their combinations. Thus, the rate of a given electrochemical reaction is generally considered to be determined according to the rate of a given elementary reaction, which is referred to as the rate-determining step. However, it is important to mention that other significant parameters including the reactant intermolecular interaction as well as adsorptive characteristics can also play a significant role. For instance, in some type of fuel cells such as hydrogen fuel cells, the rate-determining step has been suggested to be the charge transfer elementary step where adsorbed hydrogen atoms oxidize at the anode. On the other hand, in hydrocarbon fuel cells the rate-determining step is suggested to be the process of adsorption rather than the charge transfer step.

5.2.1 Activation polarization

Activation polarization corresponds to one type of voltage loss that occurs in fuel cells. This type of polarization is dependent on the activation energy associated with the rate-determining step of the electrochemical reaction. Also, this relates to the amount of resistance to the occurrence of an electrochemical reaction. Furthermore, the physiochemical steps that entail reactant adsorption also effect the activation polarization losses. As discussed earlier, the rate of a given electrochemical reaction is generally considered to be dependent on the rate-determining step. It is possible that more than one type of rate-determining step exists and the slowest reaction rate of a given elementary reaction can vary with operating conditions. However, for the purpose of analysis, it is considered that one elementary step entails the rate-limiting step, which can be written as

$$A \leftrightarrow B + n\bar{e} \tag{5.88}$$

Also, for the ease of analysis, the assumption of simultaneous oxidation as well as a reduction on the surface of the electrode is considered. The current density that results from the forward reaction can be denoted as

$$J_f = C_A K_f \tag{5.89}$$

where K_f denotes the reaction rate constant for the forward reaction and the subscript A represents the reactants. Also, similarly, the backward reaction is expressed in terms of the backward current density (J_b) as

$$J_b = C_B K_b \tag{5.90}$$

where K_b is backward reaction constant. Also, the forward current density denoted in Eq. (5.82) is also termed anodic current density as it corresponds to the oxidation reaction of a given reactant specie. More importantly, the net or overall current density of a given electrode is the difference of both types of current densities described in Eqs. (5.82) and (5.83):

$$J = J_f - J_b = C_A K_f - C_B K_b \qquad (5.91)$$

For both types of reactions (forward and backward) there exists an energy barrier that needs to be crossed for the reaction to occur and proceed. This is in accordance with the transition state theory. If the Gibbs function is plotted against the reaction completion degree for a given elementary reaction, a mountain-type curve is obtained where the intermediate species of the reaction are situated on the mountain curve top. This is often referred to as the activated complex. Moreover, it is possible for the activated complex to proceed in either way of the curve top. The reverse way corresponds to the formation of reactants and the forward way corresponds to the formation of products. Thus, for a reaction to take place the reactants need to cross the mountain curve peak. This is referred to as the energy barrier that corresponds to the change in Gibbs function between the reactant and the activated complex for the forward reaction. Also, for a backward reaction to occur the energy barrier between the Gibbs function of products and the activated complex needs to be overcome. These are also known as Gibbs functions of activation, which hold for both reaction directions. Moreover, the forward reaction rate constant can be expressed in terms of the Gibbs function of activation as

$$K_f = \exp\left(-\frac{\Delta g_{a,f}}{RT}\right) TB_f \qquad (5.92)$$

where a factor (B_f) is multiplied with the exponential term and $\Delta g_{a,f}$ denotes the Gibbs function of activation corresponding to the forward reaction. Similarly, the backward rate constant can also be represented as

$$K_b = \exp\left(-\frac{\Delta g_{a,b}}{RT}\right) TB_b \qquad (5.93)$$

where B_b represents the exponential factor that corresponds to the backward reaction and $\Delta g_{a,b}$ denotes the Gibbs function of activation corresponding to the backward reaction. In case of a reversible reaction, the fuel cell is considered to be in an open-circuit condition meaning that no net current outputs the electrode. Further, this also corresponds to the same number of electrons as well as ions crossing the electrode and electrolyte interface. This represents equal number of reactants converting to products and vice versa. Hence, both reactions (backward and forward) entail equal reaction rates. The current density corresponding to this type of situation is referred to as the exchange current density:

$$J_0 = J_b = J_f \qquad (5.94)$$

where J_0 denotes the exchange current density, J_b represent the backward reaction current density, and J_f denotes the current density relating to the forward reaction. The above equation can be rewritten and expanded as

$$J_0 = \exp\left(-\frac{\Delta g_{a,f}}{RT}\right) TB_f C_A = \exp\left(-\frac{\Delta g_{a,b}}{RT}\right) TB_b C_B \tag{5.95}$$

The exchange current density represents the activity of electron transfer corresponding to the potential of the electrode in equilibrium conditions. Furthermore, the exchange current density also provides an idea about the ease with which an electrochemical reaction occurs. Moreover, the exchange current densities associated with general fuel cell reactions entail values in the range from 10^{-18} to 10^2 A/cm². For instance, an exchange current density of 10^{-3} A/cm² is associated with a platinum electrode that is normally utilized in fuel cell applications and 10^{-12} A/cm² for a mercury electrode for the anodic reaction of hydrogen oxidation. Also, the exchange current density associated with the oxygen reduction reaction occurring at the copper electrode is 10^{-8} and 10^{-10} A/cm² at a platinum electrode. Since the exchange current density for the oxygen reduction reaction is comparatively much lower than other reactions, it is generally regarded as the fuel cell subprocess that entails higher voltage loss as compared to the anodic reaction. In hydrogen fuel cells, the oxygen reduction reaction entails the highest losses and thus significant efforts are being directed toward reducing these through the enhancement of the electrochemical reduction of oxygen. The exchange current density associated with the electrooxidation of ammonia has been reported to be approximately 10^{-8} A/cm² [70].

In case of a net current output from the cell, there are various types of irreversibilities that occur during operation. When the process is not reversible, the electrons are not transferred in a balanced way and imbalance of charge transfer is associated with the electrodes. The amount of imbalance associated with the electrode currents is signified with a net electron flow that can be in either direction (toward or away from the electrode) depending on the type of electrochemical reaction occurring at the electrode. Furthermore, this net electron flow is a function of the difference between the cell potential and the equilibrium potential that is referred to as the overpotential:

$$\eta_{op} = \epsilon - \epsilon_r \tag{5.96}$$

When the potential difference across the reactants and products at the electrode entails a value other than the reversible potential, the reaction rates of the backward, as well as forward reactions, are not equal. This results in a net flow of current. In addition, the cell overpotential aids the forward reaction to proceed by increasing the reactant energy level. Also, it reduces the product energy level to decrease the backward reaction rate. Previously, it was discussed that the change in Gibbs function can be denoted in terms of the potential difference as

$$\Delta g = -nFE \tag{5.97}$$

where the reversible cell potential is considered. In a similar manner, the change in Gibbs function considering this overpotential can also be calculated for the forward reaction as follows:

$$\Delta g_f = \Delta g_{a,f} - \eta \alpha n F \tag{5.98}$$

where the parameter α denotes the overpotential proportion that is utilized to increase the reactant Gibbs function. Also, the Gibbs function associated with the backward reaction can also be represented in a similar way as

$$\Delta g_b = \Delta g_{a,b} + \eta(1-\alpha)nF \tag{5.99}$$

where the parameter $(1-\alpha)$ denotes the amount that is utilized to decrease the product Gibbs function. Thus the reaction rate constant equations that were presented earlier in Eqs. (5.85) and (5.86) can be modified with the Δg_f and Δg_b equations given in Eqs. (5.91) and (5.92). Moreover, the current density equations for both directions of reactions can also be modified to take these into consideration. For instance, the current density equation corresponding to the forward reaction can be written as

$$J_f = \exp\left(-\frac{\Delta g_{a,f} - \eta \,\alpha n F}{RT}\right) TB_f C_A \tag{5.100}$$

Also, the backward current density can be written as

$$J_b = \exp\left(-\frac{\Delta g_{a,b} + \eta(1-\alpha)nF}{RT}\right) TB_b C_B \tag{5.101}$$

In addition to this, the above equations can be modified and written in terms of the exchange current density by substituting Eq. (5.88):

$$J_f = J_0 \exp\left(\frac{\eta \alpha n F}{RT}\right) \tag{5.102}$$

$$J_b = J_0 \exp\left(-\frac{\eta(1-\alpha)nF}{RT}\right) \tag{5.103}$$

As can be observed, overpotential at the electrode deters the backward reaction and aids the forward reaction. More importantly, from the above equations, the net current density can be evaluated at the electrode as

$$J = J_f - J_b \tag{5.104}$$

The above equation can be further expanded as

$$J = J_0 \left[\exp\left(\frac{\eta \alpha n F}{RT}\right) - \exp\left(-\frac{\eta(1-\alpha)nF}{RT}\right)\right] \tag{5.105}$$

This equation for the current density is referred to as the Butler-Volmer equation. Also, the overpotential utilized in the equation can be termed as the activation overpotential. Furthermore, the numerical parameter utilized in the equation (α) is referred to as the symmetry factor or transfer coefficient. Generally, α varies between the highest value of one and a minimum value of zero. However, several experimental studies have suggested that α entails a value of approximately 0.5. Moreover, Eq. (5.98) can also be expressed as follows:

$$J = J_0 \left[\exp\left(\frac{\eta \alpha_a nF}{RT} \right) - \exp\left(-\frac{\eta(\alpha_c)nF}{RT} \right) \right] \tag{5.106}$$

where the parameter α can be denoted corresponding to a specific electrode (anode: a or cathode: c). Also, it is important to note that since the above equation is derived from Eq. (5.98) the summation of α_a and α_c must equal to unity:

$$\alpha_c + \alpha_c = 1 \tag{5.107}$$

Once the relation between the current density and the overpotential is established, the exchange current density for a specific reactant type, products formed, and type of electrode or electrolyte can be obtained through experimental results that provide plots of $\ln(J)$ and the overpotential η. Next, further simplifications of Eq. (5.99) can be applied for special scenarios. For instance, in the case where the overpotential is considerably small, the terms inside the exponential can be assumed to entail significantly low values (lower than 0.1). In this case, the Butler-Volmer equation can be written in a simplified form as

$$J = J_0 \left\{ \frac{\eta nF}{RT} \right\} \tag{5.108}$$

Also, for the terms inside the exponential term of Eq. (5.99) to entail values of <0.1, the overpotential (η) needs to have a value of approximately lower than 0.003 V for ambient temperature operation and a 0.5 transfer coefficient with 2 mol of electrons transfers at the electrode. In addition to the case where the overpotential is considerably low, the other extreme of considerably high overpotential can also be considered. In this scenario, the terms inside the exponentials in Eq. (5.99) are considered to be higher than 1.2, which corresponds to an overpotential of higher than 0.05 V. In such cases, the Butler-Volmer equation can be expressed as

$$J = J_0 \exp\left(\frac{\eta \alpha_a nF}{RT} \right) \tag{5.109}$$

Moreover, the above equation can also be rearranged to solve for the activation overpotential as follows:

$$\eta = \ln\left(\frac{J}{J_0} \right) \frac{RT}{\alpha nF} \tag{5.110}$$

This equation is generally termed as the Tafel equation that has been used extensively in electrochemistry.

In addition to the factors considered so far, there is another parameter that needs to be accounted in the activation overpotential analysis. The exchange current density equation presented in Eq. (5.88) is a function of the concentration of reactant A (C_A) and hence can be rewritten as

$$J_0 = LC_A \tag{5.111}$$

In actual operating fuel cells, the reactant concentrations at the electrochemical interaction locations at the electrodes are normally difficult to obtain. However, the exchange current density can be obtained for a set concentration of reactant at a specified electrode, cell temperature, and electrolyte. This is referred to as the reference exchange current density:

$$J_{0,ref} = LC_{A,ref} \tag{5.112}$$

Thus, once the reference exchange current density at the set reference concentration is determined, the actual exchange current density at the actual concentration can be found as

$$J_{0,act} = J_{0,ref}\frac{C_{A,act}}{C_{A,ref}} \tag{5.113}$$

However, as in the majority of the cases, the mechanism of reactions is not always linear and include various elementary steps of reactions. In such cases, the above equation can be modified as

$$J_{0,act} = J_{0,ref}\left(\frac{C_{A,act}}{C_{A,ref}}\right)^{r} \tag{5.114}$$

where r represents the order of the reaction corresponding to the reactant A. In addition, in cases where several types of chemical species take part in the reaction, each specie concentration needs to be considered in the evaluation of the exchange current density. More importantly, the current density equation given in Eq. (5.98) can be modified to incorporate this as

$$J = J_{0,ref}\left(\frac{C_{A,act}}{C_{A,ref}}\right)^{r}\left[\exp\left(\frac{\eta \alpha nF}{RT}\right) - \exp\left(-\frac{\eta(1-\alpha)nF}{RT}\right)\right] \tag{5.115}$$

5.2.2 Polarization due to transport phenomena

When a fuel cell operates, parallel transport of electricity, momentum, heat, and mass take place. Whether the transport and transfer mechanisms take place in a fast or slow rate significantly effects the output fuel cell performance. The reactant molecules flow in the anodic compartment and the product molecules flow out through the cathode compartment and the electrochemical interactions within the fuel cell generate electrical power. As various simultaneous processes keep occurring within the fuel cell, different species interact in an interlinked manner in different phases in various dimensions. This increases the complexity of mathematical formulation for these phenomena occurring within the fuel cell. In this subsection, an effort is made to briefly describe the transport processes in fuel cells and the resulting concentration polarization losses that occur in the cell. For a mixture entailing different species the concentration of one of the included specie can be denoted as

$$C_{n,A} = \frac{n_A}{V_t} \tag{5.116}$$

where $C_{n,A}$ represents the molar concentration of specie A, the total volume is denoted by V_t, and the number of moles of specie A is represented by n_a. Similarly, the concentration of specie A on a mass basis can be denoted as

$$C_{m,A} = \frac{m_A}{V_t} \tag{5.117}$$

where $C_{m,A}$ is the mass concentration of specie A (in kg/m^3) that entails the mass of m_A in a total mixture volume of V_t. As can be observed, this entails the same units of density and is thus generally known as partial density. Utilizing the molar mass of the given specie, the partial density can also be written as

$$C_{m,A} = C_{n,A} M_A \tag{5.118}$$

where M_A represents the molar mass of specie A. In addition to this, the ratio of the number of moles of specie A with respect to the total number of moles in the mixture is known as the mole fraction that can be expressed as

$$F_A = \frac{n_A}{N_t} \tag{5.119}$$

where F_A denotes the mole fraction, n_A is the number of moles of specie A, and N_t denotes the total number of moles in the mixture. This can also be expressed in terms of the molar concentration as

$$F_A = \frac{C_{n,A}}{C_t} \tag{5.120}$$

where C_t denotes the total concentration on a molar basis considering all species in the mixture. Similarly, the mass fraction can also be written as

$$MF_A = \frac{m_A}{m_t} \tag{5.121}$$

where m_A is the mass of specie A and m_t is the total mixture mass. The above equation can also be rewritten as

$$MF_A = \frac{C_{m,A}}{C_{m,t}} \tag{5.122}$$

where $C_{m,t}$ denotes the summation of the mass concentration of all species that is also known as the density of the mixture. Considering these equations, the summation of the mole and mass fractions of all species should equal to unity:

$$\sum_{a=1}^{N} F_A = 1 \text{ and } \sum_{a=1}^{N} MF_A = 1 \tag{5.123}$$

Thus, the molecular mass of a given mixture can be denoted in terms of the molar fractions (F_A) and molecular masses (M_A) of each specie as

$$M_{t,mix} = \sum_{a=1}^{N} F_A M_A \tag{5.124}$$

Also, using the above definition both mass and molar fractions can be related:

$$MF_A = F_A \frac{M_A}{M_{t,mix}} \tag{5.125}$$

Next, as fuel cell reactant and product flows entail certain velocities, it is important to consider the velocities of mixture. In a given mixture, if the absolute velocity of given specie A is denoted by v_a, the average velocity of the mixture can be evaluated on a mass basis as

$$v_{avg,mix} = \frac{\sum_{a=1}^{N} C_{m,A} v_a}{\sum_{a=1}^{N} C_{m,A}} \tag{5.126}$$

The above equation can also be written in terms of the mass fraction as

$$v_{avg,mix} = \sum_{a=1}^{N} MF_A v_A \tag{5.127}$$

Similarly, the average velocity can also be written in terms of the molar fractions that can be expressed as

$$v_{avg,mix,m} = \sum_{a=1}^{N} F_A v_A \tag{5.128}$$

Further, the velocity of specie A that denotes the diffusion velocity can be expressed on a mass basis as

$$v_{A,diff} = v_A - v_{avg,mix} \tag{5.129}$$

Also, the above equation can be written on a mole basis according to

$$v_{A,diff,m} = v_A - v_{avg,mix,m} \tag{5.130}$$

Thus, when a mixture entails an average velocity, a given specie A diffuses with a velocity of $v_{A,diff}$ that depends on both the absolute velocity of the specie as well as the average velocity of the mixture. A typical example of the application of these is the flow of air in the cathodic compartment of ammonia fuel cells. The oxygen entailed in the air represents one type of specie that undergoes diffusion.

Once the velocities are discussed, the mass transfer rates can be described. For a given specie A entailed in the mixture, the mass transfer rate can be expressed as

$$\dot{m}_A'' = C_{m,A} v_A \tag{5.131}$$

where $\dot{m}_A''$ denotes the mass flow flux of specie A where the mass flow rate per unit normal area per unit time is presented. The normal area denotes the direction normal to the flow of the species. Similarly, the molar flux of specie A can also be written as follows:

$$\dot{N}_A'' = C_{n,A} v_A \tag{5.132}$$

where $\dot{N}_A''$ denotes the molar flux of specie A in the normal direction. In addition, the mass flux can also be denoted in terms of the relative fluid motion at the average value as

$$\dot{m}_{A,d}'' = C_{m,A} v_{A,\,diff} \tag{5.133}$$

Also, the molar flux can be written as follows:

$$\dot{N}_A'' = C_{n,A} v_{A,\,diff,\,m} \tag{5.134}$$

where $\dot{m}_{A,d}''$ and $\dot{N}_A''$ denote the diffusion fluxes on a mass and molar basis, respectively.

Moreover, the total mass flux per unit time of the complete mixture including all species can be denoted as

$$\dot{m}'' = \sum_{a=1}^{N} \dot{m}_A'' \tag{5.135}$$

Similarly, the total molar flux of the mixture is represented as

$$\dot{N}'' = \sum_{a=1}^{N} \dot{N}_A'' \tag{5.136}$$

Since fuel cell operation includes the diffusion of a given specie in a mixture of species, the Fick's law of diffusion is normally utilized to study the diffusion mechanism. It can be expressed as

$$\dot{m}_{A,d}'' = -\rho D_{A,B} \nabla MF_A \tag{5.137}$$

where the mass transfer flux rate is denoted by $\dot{m}_{A,d}''$, ρ is density, $D_{A,B}$ represents the coefficient of diffusion for specie A with respect to B. This is also referred to as the binary mass diffusivity. Moreover, as can be observed from the equation, the mass flux diffusion of specie A results due to the gradient in the mass concentration. Also, as can be depicted from the negative sign, the direction of diffusion results in the decreasing concentration direction. Stated in simpler terms, diffusion occurs from the high concentration region of specie A to the low concentration region. Moreover, Eqs. (5.126) and (5.130) can be combined to obtain the following relation:

$$MF_A v_{A,\,diff} = -D_{A,B} \nabla MF_A \tag{5.138}$$

Thus, from the above equation, the specie A velocity of diffusion can be evaluated when the mass fractions are determined. Moreover, when a specie diffuses, the total mass flux results due to both the gradient in the low and high concentration as well as the bulk mixture velocity. Similar to the mass flux equations, the molar flux equation can also be written as

$$\dot{N}_{A,d}'' = -C_{n,A} D_{A,B} \nabla F_A \tag{5.139}$$

Also, Eqs. (5.127) and (5.132) can be combined to form the following relation:

$$C_{n,A} v_{A,\,diff,\,m} = -D_{A,B} \nabla F_A \tag{5.140}$$

After presenting the concepts of mass diffusion, we describe the next type of polarization loss corresponding to the concentration polarization. In the discussion of the activation polarization, the specie concentrations were considered constant and did not vary with the current. However, in actual fuel cell operation, there exist concentration gradients that entail high concentrations near the region where there are the bulk electrolyte amount and low concentrations near the surface of the electrode. The presence of a low number of reactant molecules near the electrode surface and the increasing number of molecules away from the vicinity of the electrode leads to a polarization loss, which is reflected in the form of a voltage loss in the operating cell. The presence of a low number of reactant molecules near the surface of the electrode can be attributed to the limitations in the mass transfer of participating chemical species. Moreover, when a higher amount of current is extracted from the cell, a higher number of electrochemical reactions need to take place to compensate for the increased current output. As the electrochemical activity rises, the number of molecules consumed in the electrochemical reaction also increases. However, owing to mass transfer limitations the number of molecules reaching the electrode surface may not be sufficient to meet the high current output requirement. Thus, there exists an upper limit on the amount of current that can be extracted from a given fuel cell. Beyond this limit, the limitation in transfer and diffusion rates does not allow the sufficient number of molecules to reach the reaction sites and participate in the electrochemical reactions to meet the increased current. The maximum current density that can be attained is referred to as the limiting current density. In the vicinity and beyond this value of current density, the cell output voltage is considerably reduced due to significant losses that can even result in no useful output voltage. As described earlier, the limiting current density is a function of the mass transport limitations in the cell. These include various subprocesses that occur within the cell. When the reactants and products flow through the embedded channels for flow in the flow channel plates, it is referred to as the convective mass transfer. Next, the process of molecular diffusion into the pores of electrodes occurs. Furthermore, the dissolution or solution of the participating species out or into the electrolyte happens. Moreover, the reactant molecules diffuse through the electrolyte to reach the other electrode surface and the product molecules diffuse away from the reaction sites. Thus, primarily low diffusion coefficients or solubility can be identified as the main contributors to the concentration polarization as these inhibit the molecules from reaching the reaction sites or moving away from them. However, it should be noted that concentration polarization is a phenomenon that occurs at high current outputs. The value of the limiting current density depends on various factors including the type of electrode, electrolyte, specie diffusivity, etc. In addition to this, concentration polarization can also be caused due to an accumulation of products in the vicinity of active sites, which prevent new unreacted molecules from reacting electrochemically as they are not able to reach the sites suitable for the electrochemical reactions. For instance, when an alkaline direct ammonia fuel cell is operated the products include the formation of water molecules. As the operating temperature is generally low, the water molecules entail a liquid phase. If these molecules do not leave the electrode

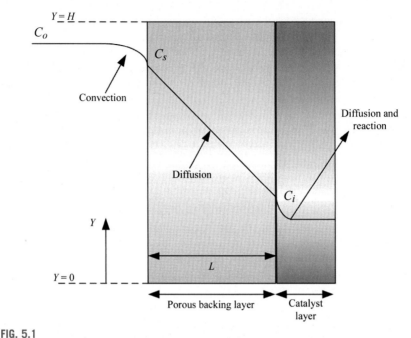

FIG. 5.1

Schematic depicting the classical approach toward the mass transport analysis.

surface and exit the cell, other new reactant molecules will be inhibited from reaching the appropriate reaction locations. This can lead to significant deterioration of cell performance. To analyze the reactant molecular transport within the cell, there are several methods. The classical approach entails the consideration of the electrode as depicted in Fig. 5.1. The electrode, in this case, is considered to include a backing layer with porous properties and a catalyst layer. As can be observed, the porous layer entails much higher thickness than the catalyst layer. Further, before the porous layer, the molecular concentration C_0 represents the concentration at the respective compartment (anodic or cathodic) before the molecules enter the porous layer. The reactant molecules move from the bulk concentration area to the surface of the electrode primarily by mass convection. The rate of convection of the reactant molecules can be denoted as

$$\dot{N}''_{conv} = h_{conv}(C_0 - C_s) \qquad (5.141)$$

where $\dot{N}''_{conv}$ denotes the transfer rate due to convection per unit surface area of electrode, h_{conv} represents the coefficient of mass transfer, C_0 is the reactant concentration in the bulk region, and C_s is the reactant concentration at the surface of the electrode. In addition to this, the transfer rate according to Fick's law through the porous layer between the surface concentration (C_s) and the concentration at the interface of the porous layer and the catalyst layer (C_i) can also be denoted as

$$\dot{N}''_{conv} = D^{eff} \frac{C_s - C_i}{L} \tag{5.142}$$

where D^{eff} denotes the effective diffusion coefficient and the thickness of the porous layer is denoted with L. Moreover, the mass transfer resistances are denoted as follows:

$$R_{conv} = \frac{1}{h_{conv}} \tag{5.143}$$

$$R_{diff} = \frac{L}{D^{eff}} \tag{5.144}$$

Furthermore, the transfer rate incorporating both diffusion and convection can be written as follows:

$$\dot{N}'' = \frac{C_0 - C_i}{R_{conv} + R_{diff}} \tag{5.145}$$

where the convective mass transfer resistance is denoted by R_{conv} that is expressed by Eq. (5.136) and the diffusion mass transfer resistance is denoted by R_{diff} that is evaluated through Eq. (5.137). Moreover, the effective diffusion coefficient (D^{eff}) is dependent on several factors including the diffusion mechanism of the bulk reactant side as well as the structure of porous layer. If the diffusion coefficient in the bulk region is denoted by D, the effective diffusion coefficient can also be expressed as follows:

$$D^{eff} = D\varphi^{1.5} \tag{5.146}$$

where φ denotes the porosity of the layer considering uniform properties. Also, the exponent factor of 1.5 is applied in accordance with the Bruggenman correction type, which is derived from empirical data. Further, the diffusion coefficient in the bulk region (D) varies with changing operational parameters including operating pressure, temperature, type of reactants, and their respective molecular dimensions. In addition, the molar transfer rate of reactants can also be related to the current density (J) according to Faraday's law as

$$\dot{N}'' = \frac{J}{nF} \tag{5.147}$$

Further, Eq. (5.140) can be utilized along with Eq. (5.138) to obtain

$$\frac{J}{nF} = \frac{1}{\frac{1}{h_{conv}} + \frac{L}{D^{eff}}} (C_0 - C_i) \tag{5.148}$$

As can be observed from the above equation, the amount of current density will be proportional to the difference between the concentration at the bulk region (C_0) and the concentration at the interface (C_i). Also, if a higher concentration of reactant is used in the bulk region (C_0) then a higher current density can be obtained. In addition, if the interface concentration (C_i) decreases, the current density attains a high value.

This corresponds to a higher amount of reactant interaction that leads to lower concentrations at the interface. In the case where the interface concentration approaches zero, the maximum amount of current density possible can be written as

$$\frac{J_L}{nF} = \frac{1}{\frac{1}{h_{conv}} + \frac{L}{D^{eff}}}(C_0 - C_i) \tag{5.149}$$

where J_L represents the limiting current density. As can be observed from the above equation, the limiting current density varies with changing operating pressures as well as temperatures. Also, the thickness of the porous layer, as well as its porosity, play a vital role in determining the limiting current density. The limiting current density is generally convenient to measure through experimental investigations. Moreover, if we combine Eqs. (5.141) and (5.142), we can express the relation between the current density and the limiting current density in terms of the concentrations as

$$\frac{J}{J_L} = 1 - \frac{C_i}{C_0} \tag{5.150}$$

rearranging the above equation we can obtain the expression:

$$\frac{C_0}{C_i} = \frac{J_L}{J_L - J} \tag{5.151}$$

Furthermore, recall that expression of the Nernst potential can be written as

$$E_{rev}(P_i, T) = E_{rev}(P, T) - \frac{RT}{nF} \ln \prod_{a=1}^{N} \left(\frac{P_a}{P}\right)^{\frac{\gamma_a'' - \gamma_a'}{\gamma_f'}} \tag{5.152}$$

Also, the ideal gas law for mixtures can be used to relate the partial pressures with the concentrations as follows:

$$\frac{C_a}{C} = \frac{P_a}{P} \tag{5.153}$$

The above equation can be substituted into Eq. (5.145) to obtain an expression for the Nernst potential in terms of the concentrations. Further, consider that all the reactant molecules interact at the interface between the catalyst and the porous layer. Also, assume the effects of desorption and adsorption phenomena, as well as other surface interactions, have insignificant effects on the voltage loss. With these assumptions, the concentration polarization is expressed as

$$\eta_{con} = \frac{RT}{nF} \ln \prod_{a=1}^{N} \left(\frac{C_0}{C_i}\right)^{\frac{\gamma_a' - \gamma_a''}{\gamma_f'}} \tag{5.154}$$

Next, the above equation can be rewritten by substituting Eq. (5.144) as

$$\eta_{con} = \frac{RT}{nF} \ln \prod_{a=1}^{N} \left(\frac{J_L}{J_L - J}\right)^{\frac{\gamma_a' - \gamma_a''}{\gamma_f'}} \tag{5.155}$$

If the consumption of reactants is assumed to be complete, the term γ_a' approaches zero. In this case, the concentration polarization can be written as

$$\eta_{con} = \frac{RT}{nF} \ln \left(\frac{J_L}{J_L - J} \right) \qquad (5.156)$$

However, the above equation is valid under the assumption that the reaction occurs only at the interface. This can be considered in case of high current densities, nevertheless, when the currents are low the entire catalyst layer may be utilized. Furthermore, the above equation is valid for the assumption that the low rate of specie transport is the limiting step of the overall process. This assumption becomes valid only when the output currents are significantly high. Thus, the above equation can be utilized if these assumptions represent a given operating fuel cell system. As mentioned earlier, there are other approaches that have been investigated for evaluating the concentration polarization voltage losses. Here only the classical approach has been covered and other methods can be found in available fuel cell literature.

5.2.3 Polarization due to electricity transport

The next type of polarization loss that occurs in fuel cells entails the voltage losses owing to the presence of electrical resistance within the system. When electrons are emitted at the anode, an external circuit connection with the cathodic side of the fuel cell generates an electric current that passes through the external load. This flow of electrons occurs through a series of fuel cell components that entail Ohmic resistances. For instance, when the electrons emitted pass through the porous layer, the resistance to electron flow through the layer causes voltage losses. Also, resistance to electron flow at the interfaces between each component of the cell also leads to losses in voltage. Moreover, the electrical components included in the external circuit also entail electrical resistances. These also lead to voltage losses and the amount of losses are proportional to the electrical current drawn from the cell. In addition to electronic resistances of various cell components, there exist ionic resistances. This includes the resistance of liquid or membrane electrolyte to the flow of ions. As discussed earlier, the fuel cell electrolyte entails the function of transferring ions from one electrode to other. In anion exchange membrane-based direct ammonia fuel cells, the electrolyte transfers negatively charged OH^- ions through the membrane electrolyte from the cathodic side of the cell to the anodic side. In PEM fuel cells, the proton exchange membrane electrolyte allows the transfer of positively charged H^+ through the electrolyte from the anode to the cathode. Thus, the resistances associated with electrolytes result in considerable voltage losses. As the amount of current drawn from the cell increases, the amount of electron transfer rises. This results in higher Ohmic losses across the cell at higher currents. The voltage loss occurring from the total resistances within the cell can be written as

$$V_{Ohm} = J R_{tot} \qquad (5.157)$$

where J is the current drawn from the cell and R_{tot} is the total Ohmic resistance of the cell including all components such as electrolyte, electrode, connection wirings, etc. In fuel cells that are developed appropriately, the primary source of Ohmic resistance is often identified as the electrolyte. The resistance of a given component can be expressed as

$$R = \tau \frac{\delta}{A} \tag{5.158}$$

where τ denotes the resistivity, A represents the cross-sectional area, and δ denotes the conducting path length. The area utilized in the above equation is the area that is normal to the current flow. Also, resistivity is a property of the material utilized and varies according to the type of material used. Furthermore, often the resistivity is expressed in terms of the area specific resistance (τ_A) that can be written in terms of the resistivity and the conducting path length as

$$\tau_A = \tau \delta \tag{5.159}$$

In addition to this, as resistivity is a property of the material, the conductance of a given material can also be expressed in terms of the resistivity as

$$C = \frac{1}{\tau} \tag{5.160}$$

Thus, Eq. (5.151) can be rewritten in terms of the conductance as follows:

$$R = \frac{1}{C} \frac{\delta}{A} \tag{5.161}$$

Furthermore, the voltage loss due to Ohmic resistance can thus be expressed as

$$V_{Ohm} = JA \frac{1}{C} \frac{\delta}{A} \tag{5.162}$$

It should be noted that the area A is often taken as the active area of the electrode. Also, as can be observed from the above equation the Ohmic voltage loss can be decreased through the usage of lower thickness electrolytes. Furthermore, electrolytes entailing higher conductance will lead to lower Ohmic voltage losses. However, using lower thickness electrodes may not always aid in reducing the Ohmic voltage losses. Appropriate flow and transport of react molecules as well as electrical transport are important factors that depend on the electrode thickness. Hence, the usage of suitable electrodes is essential that can provide both appropriate specie transport as well as lower Ohmic voltage losses.

5.2.4 Performance assessment

The performance of a given type of fuel cell is assessed through different performance indicators. First, the open-circuit voltage is a system parameter that reflects how closely the electrochemical interactions occurring in the cell resemble the theoretical performance. The reversible voltage under open-circuit conditions discussed earlier in Eq. (5.23) denotes the cell voltage under ideal operation where no irreversibilities cause entropy generation or exergy destruction. However, in actual fuel cell operation, the open-circuit voltage is below this reversible value. Thus, one way to assess the fuel cell performance entails the comparison of the voltage under open-circuit conditions with the theoretical voltage. Closer the obtained voltages are to the theoretical value, higher is the fuel cell performance. The open-circuit voltages obtained through different types of direct ammonia fuel cells have been discussed in Chapter 3. Furthermore, the performance of a given fuel cell is also assessed through its peak power density. During the operation of a given fuel cell, power output density increases with increasing current density until it reaches its maximum value. This is known as the peak power density, where increasing the current density further than this value decreases the output power. This is an important performance assessment parameter that is the desired operating point during practical fuel cell usage and applications. The system design is conducted in such a way that the operating current density and cell voltage are near the peak power density point. Moreover, higher the peak power density, lower is the electrode active area required to attain a given power output. Also, higher the peak power density, higher is power output for the same electrode active area. In addition to this, the short-circuit current density is another parameter that aids in assessing the performance of a fuel cell. As the current density increases, the cell voltage decreases due to various polarization losses as discussed earlier. Thus, the maximum current that can be achieved from a given cell can be used to assess the significance of voltage losses. In addition, the short-circuit current density also provides a maximum limit for fuel cell operators to know when how much current can be extracted from the cell. Higher the short-circuit current density, lower are the overall voltage losses in the cell. Furthermore, energy and exergy efficiencies are one of the most important performance assessment parameters. These describe the ratio of the useful electrical power output from the cell to the energy input to the cell in terms of the lower or HHV of the fuel utilized.

The energy efficiency relation for an ammonia fuel cell can be expressed on a LHV basis as

$$\eta_{AFC,en,LHV} = \frac{\dot{P}_{out}}{\dot{N}_{i,NH_3}\overline{LHV}_{NH_3}} \tag{5.163}$$

Also, the energy efficiency for an ammonia fuel cell can be expressed on a HHV basis as

$$\eta_{AFC,en,HHV} = \frac{\dot{P}_{out}}{\dot{N}_{i,NH_3}\overline{HHV}_{NH_3}} \tag{5.164}$$

where the usage of LHV or HHV is dependent on the formation of water in the liquid or vapor state. The parameter $\dot{P}_{out}$ denotes the power output from the ammonia fuel cell, $\dot{N}_{NH_3}$ is the molar input flow rate to the fuel cell, $\overline{LHV}_{NH_3}$ is the molar LHV of ammonia and $\overline{HHV}_{NH_3}$ is the molar HHV of ammonia. In addition, the exergy efficiency of an ammonia fuel cell can be expressed as

$$\eta_{AFC,ex} = \frac{\dot{P}_{out}}{\dot{N}_{i,NH_3}\overline{ex}_{tot,NH_3}} \tag{5.165}$$

where the total specific molar exergy of ammonia is denoted by $\overline{ex}_{tot,NH_3}$ that can be evaluated as

$$\overline{ex}_{tot,\mathrm{NH_3}} = \overline{ex}_{phy,\mathrm{NH_3}} + \overline{ex}_{ch,\mathrm{NH_3}} \qquad (5.166)$$

Another methodology of determining the efficiencies of ammonia fuel cells includes considering the molar rate of ammonia consumption during the electrochemical interactions. This can be evaluated as the difference between the inlet and exit molar flow rates considering no mass accumulation:

$$\dot{N}_{con,\mathrm{NH_3}} = \dot{N}_{i,\mathrm{NH_3}} - \dot{N}_{o,\mathrm{NH_3}} \qquad (5.167)$$

where the subscripts i and o denote inlet and outlet, respectively. In this case, the energy efficiencies can be expressed as

$$\eta_{AFC,en,LHV} = \frac{\dot{P}_{out}}{\dot{N}_{con,\mathrm{NH_3}}\overline{LHV}_{\mathrm{NH_3}}} \qquad (5.168)$$

$$\eta_{AFC,en,HHV} = \frac{\dot{P}_{out}}{\dot{N}_{con,\mathrm{NH_3}}\overline{HHV}_{\mathrm{NH_3}}} \qquad (5.169)$$

Similarly, the exergy efficiency can be written as follows:

$$\eta_{AFC,ex} = \frac{\dot{P}_{out}}{\dot{N}_{con,\mathrm{NH_3}}\overline{ex}_{tot,\mathrm{NH_3}}} \qquad (5.170)$$

However, it is important to note that these equations are applicable to systems that do not include fuel recycling. For systems where the unreacted fuel is recycled, the total fuel inlet to the cell is generally applied in the efficiency calculations. Also, the utilization factor can be applied to the efficiency equations to account for unreacted fuel.

An exemplary figure depicting the performance assessment parameters of open-circuit voltage, short-circuit current density, and peak power density is shown in Fig. 5.2. These results depict the modeling results for an anion exchange membrane direct ammonia fuel cell operating at 25°C. The simulation parameters including the exchange current density, membrane type, Ohmic resistance, etc., are acquired from the experimental results conducted at CERL. As can be observed from Fig. 5.2, the open-circuit voltage is 0.35 V. Also, the peak power density is found to be 6.49 W/m^2 and the short-circuit current density is 72 A/m^2. Hence, the open-circuit voltage obtained (0.35 V) is lower than the theoretical voltage of 1.17 V. This entails a difference of 0.82 V between the theoretical and actual value. Thus, the first performance improvement potential for ammonia fuel cells includes the reduction of this voltage difference between actual and theoretical values. Lower the voltage difference, higher will be the overall fuel cell performance.

Furthermore, the peak power density of 6.49 W/m^2 is another important performance assessment parameter. The improvement potential of ammonia fuel cells also entails the enhancement in the peak power densities. Higher peak power densities will result in higher useful output from the fuel cell and will thus result in both higher energetic and exergetic efficiencies. The peak power density for ammonia fuel cells entails a comparatively lower value and efforts can be focused

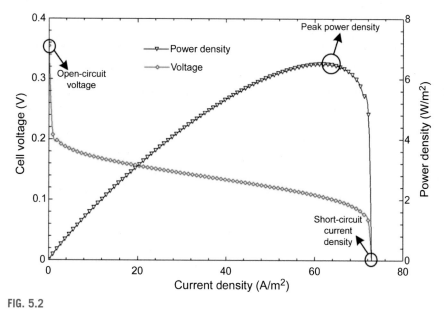

FIG. 5.2

Performance assessment parameters of open-circuit voltage, short-circuit current density, and peak power density.

toward its improvement. In addition, the short-circuit current density is also a parameter that allows an assessment of the maximum attainable current from the fuel cell. Hence, higher the short-circuit current density, better will be the overall fuel cell performance. Moreover, this can be directly related to the polarization losses as discussed earlier. Higher the polarization losses, lower is the short-circuit current density. Thus, the overall performances of ammonia fuel cells can be enhanced through the reduction of polarization losses. These include activation polarization, Ohmic polarization, and concentration polarization.

5.3 Closing remarks

In this chapter, the fundamental thermodynamic concepts required to comprehend the thermodynamic as well as electrochemical modeling of ammonia fuel cells is firstly described. The evaluation of the reversible cell potential through the usage of these fundamental concepts is presented and its underlying significance is discussed. Further, Gibbs's free energy concept is described in detail and its relation to the cell voltage is presented. Next, the usage of the Nernst equation to determine the reversible cell potentials is explained depicting the effects of varying temperatures and pressures on these voltages. Moreover, the adsorption phenomena of ammonia molecules at the active sites of the electrochemical catalysts are described.

The elementary steps involved in the dissociation of nitrogen and hydrogen bonds at the actives sites and corresponding electron release is discussed. In addition to this, the activation polarization phenomenon is comprehensively described elucidating the concept of exchange current density and the voltage loss that occurs during the initial reaction mechanism. Also, the concentration polarization losses are described explaining the effects of molecular diffusion on the loss in voltages at high current densities. Further, the voltage losses due to electrical resistances are also covered. Lastly, the method of assessing the performance of ammonia fuel cells through open-circuit voltages, peak power densities, evaluation of energy, and exergy efficiencies is described.

Integrated ammonia fuel cell systems

Ammonia fuel cells (AFCs) can be implemented in renewable energy-based integrated systems. Such systems provide the opportunity to utilize the electrochemical interactions of ammonia molecules for the production of useful output commodities through the operation of an integrated system. Several integrated AFC systems have been developed that will be described in the proceeding sections along with discussions about their thermodynamic performances at varying operating parameters and system conditions.

6.1 Integrated AFC and thermal energy storage system

An integrated alkaline molten salt-based hybrid thermal energy storage (TES) and direct ammonia fuel cell (DAFC) system is depicted in Fig. 6.1. The system entails hybrid operations of storing and discharging thermal energy as well as providing electrical power output though electrochemical interactions of ammonia fuel in the molten alkaline electrolyte. The anodic interactions as depicted in the figure include the electrochemical reaction of ammonia molecule with the hydroxyl ions in the molten electrolyte to form nitrogen, water, and free electrons. Simultaneously, the oxygen and water molecules interact electrochemically at the cathode where they accept the electrons incoming from the external circuit to complete the overall reaction.

Such systems can be implemented in various applications including solar thermal power plants as well as waste heat recovery systems. The electrolyte comprises molten potassium hydroxide (KOH) and sodium hydroxide (NaOH) salts with a mass ratio of 1:1. Ammonia fuel is input to the system as a gas along with humidified air or oxygen. The humidified air or oxygen produces hydroxyl ions electrochemically in the presence of a catalyst that recycles the usage of hydroxyl ions at the anode. Thus, in this way, the molten salt TES system entailing the storage of excess or waste thermal energy provides an innovative way to produce clean power from ammonia electrochemically. The depicted system has been developed and tested at the clean energy research laboratory. A ceramic alumina flask is used as the containment holding the molten salt mixture within an insulated container that can withstand high temperatures. The input reactant passages comprise stainless steel tubing that allows effective flow of gases. The flow tubing is attached to nickel

Ammonia Fuel Cells. https://doi.org/10.1016/B978-0-12-822825-8.00006-2

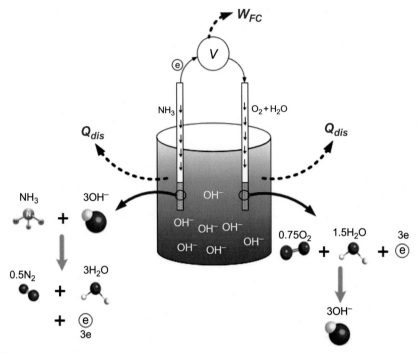

FIG. 6.1

Schematic representation of the integrated molten salt-based ammonia fuel and thermal energy storage system.

coil electrodes immersed in the molten salt containment. Nickel has been identified as one of the appropriate electrochemical catalysts that do not entail activities as high as platinum but is associated with lower costs. The total electrode area of the coils was measured to be $2.4 \pm 0.5\,\text{cm}^2$. The ammonia fuel input at the anode was set at $0.2\,\text{mg/s}$ to ensure no splashing of high-temperature molten salt. The inlet pressure of ammonia was set at $101.325\,\text{kPa}$. Further, the humidification needed for air input at the cathode was obtained via a bubbler humidifier described earlier. The performance of the experimental system developed is investigated at varying operating temperatures. The electrochemical performance of the molten salt-based DAFC investigated experimentally is depicted in Figs. 4.21 and 4.22 in Chapter 4. As can be observed, higher temperatures result in better fuel cell performances. Higher temperatures result in higher operable current densities that provide higher peak power outputs. However, the open-circuit voltage reduces with rising operating temperatures. The highest peak power density is observed at an operating temperature of 320°C. The developed molten salt-based system is investigated electrochemically at temperatures of 220, 270, and 320°C. The energy efficiency of the fuel cell operation is evaluated from

$$\eta_{MFC} = \frac{P_d A_E}{\overline{LHV}_{NH_3} \dot{n}_{NH_3}} \tag{6.1}$$

where the power density is denoted with P_d, the area of the electrodes is represented with A_E, the lower heating value of ammonia is denoted with $\overline{LHV}_{NH_3}$ on a molar basis, and $\dot{n}_{NH_3}$ denotes the consumption rate of input ammonia gas. In this regard, the exergetic performance of the fuel cell operation is determined from

$$\eta_{MFC} = \frac{P_d A_E}{\overline{ex}_{NH_3} \dot{n}_{NH_3}} \tag{6.2}$$

where the denominator includes the specific exergy content of ammonia molecules on a per unit mole basis ($\overline{ex}_{NH_3}$). Moreover, the consumption rate of ammonia molecules is determined from the current density (J_{MFC}), electrode area (A_E), number of electrons (n_e), and Faraday's constant (F) as

$$\dot{n}_{NH_3} = \frac{J_{MFC} A_E}{n_e F} \tag{6.3}$$

The capacity of storing thermal energy for the developed system is also analyzed on a per unit mass basis as

$$q_{MFC} = c_{p,st}(T_{MFC} - T_i) \tag{6.4}$$

where the total energy storage capacity is represented by q_{MFC}, the specific heat capacity is represented by $c_{p,st}$, the initial temperature is denoted by T_i that is set at 220°C. The energy efficiency results are depicted in Fig. 6.2A for varying operating temperatures as well as current densities. Also, Fig. 6.2B depicts the exergy efficiency results for these parameters. The efficiencies at the peak power density point are observed to entail values in the vicinity of 20%–23% where at varying temperatures, the efficiency values entail similar values near the peak power density range.

The results of the specific heat capacity of the system at varying operating temperatures is shown in Fig. 6.3. The heat capacity of the system at 320°C is evaluated to be 133 kJ/kg of molten alkaline electrolyte salt. This reduces with decreasing temperatures with a specific heat capacity of 13.3 kJ/kg at 230°C considering an initial operating temperature of 220°C. Thus, the developed system can be implemented in various applications where excess or waste thermal energy is available, for both thermal storage as well as electrochemical DAFC operation. It is recommended to investigate the changes in the thermos-physical properties of the molten alkaline salt when passed through a fuel cell chamber or when ammonia fuel directly inputs to the system. Depending on the temperature of the gas input, the temperature inside the tank may decrease due to heat transfer between the salt and the reactant gases. Also, the molten salt can be circulated through external fuel cell electrodes where the reactant gases have their respective flow channels and are not trapped in the molten salt medium.

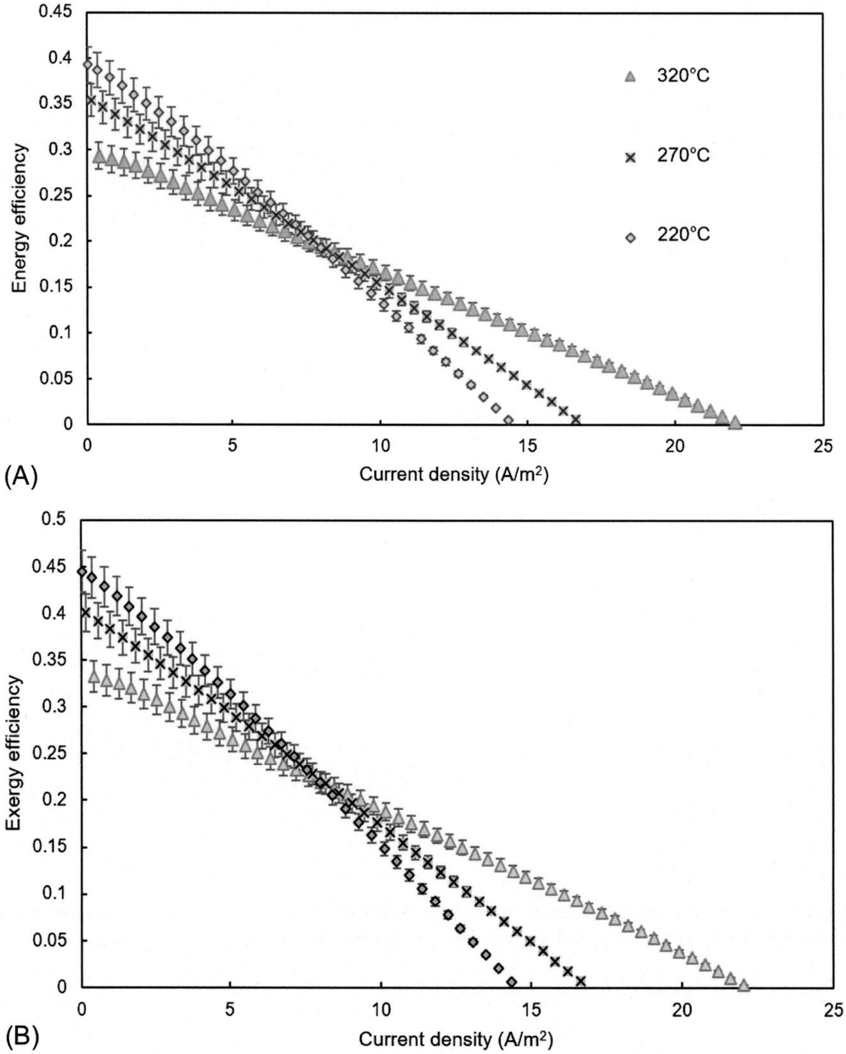

FIG. 6.2

Efficiency results for the molten salt-based ammonia fuel cell and thermal energy storage system showing (A) energy efficiency and (B) exergy efficiency.

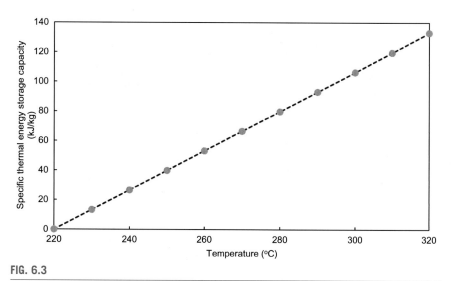

FIG. 6.3

Results for specific thermal energy storage capacity of the molten electrolyte-based hybrid AFC and thermal energy storage system.

6.2 DAFC integrated with solar thermal power plant

The developed solar thermal power plant integrated with the hybrid TES and DAFC systems is shown in Fig. 6.4. The heliostat field reflects incoming solar radiation onto the central tower receiver where the molten alkaline salt is passed to absorb the incoming thermal energy and attain a higher temperature according to the received energy. The high-temperature molten salt is sent to the hybrid system after absorbing incoming solar energy. A tank reservoir-based system as depicted in the figure is considered in the present study. The alkaline molten electrolyte is passed through heat exchanger HX-1 where it transfers thermal energy to the multistage reheat Rankine cycle for power generation. At state 2, water enters HX-1 at the highest cycle pressure where it absorbs the thermal energy from the molten salt and converts to superheated vapor at state 3 before entering the multistage turbine (MSRT).

Moreover, after leaving the high-pressure turbine at state 4, the intermediate temperature and pressure steam are reheated by sending it back to HX-1. At state 5, it enters the intermediate pressure turbine and is reheated again to state 7 after exiting at state 6 before entering the low-pressure turbine. At state 8, saturated steam exits the low-pressure turbine and enters condenser (CON) to reject heat before being pumped to the highest cycle pressure. This steam Rankine cycle operates according to the thermal energy available from the sun. These vary across the year and are dependent on the location, time, and weather. The TES system is employed in such solar plants to mitigate the intermittency issues through the storage of excess

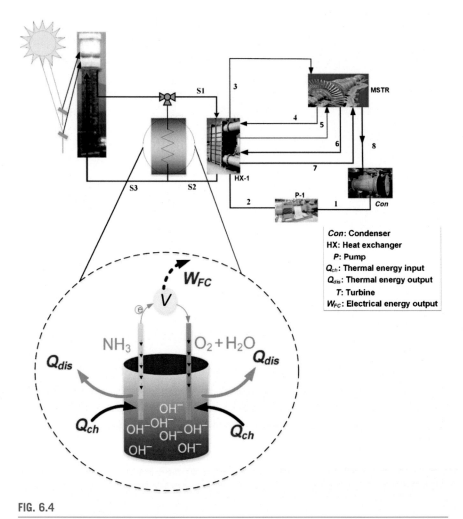

FIG. 6.4

Schematic representation depicting the solar thermal power plant integrated with hybrid thermal energy storage and direct AFC system.

available solar energy. This is described as the charging phase where the excess solar thermal energy is stored in a molten salt tank. Until the stored energy is discharged, it is stored in the insulated tank which is referred to as the storing phase. The stored energy is later discharged through HX-1 where thermal energy is transferred to the reheat Rankine cycle for power generation. This is referred to as the discharging phase. When solar radiation intensities are available but are not in excess, the power plant is operated under normal operation and no excess energy storage occurs. In the developed system, during the discharging phase, the discharge of both thermal, as well as electrical energy, occurs through the hybrid TES+DAFC system. The molten alkaline salt is sent to HX-1 to transfer heat to the reheat Rankine cycle and simultaneously the alkaline salt is used as an electrolyte to generate electrical power through the DAFC operation. Thus, when the low availability of solar energy occurs, the stored thermal energy is used along with ammonia fuel to generate electrical power. The thermal energy is transferred to the power generation Rankine cycle whereas the ammonia fuel interacts electrochemically through the passage of hydroxyl ions through the same molten electrolyte that is used for TES.

Each system component is analyzed thermodynamically to determine the overall performance. The general mass balance equation applied to each component of the system can be written as

$$\sum_{in} \dot{m}_{in} = \frac{dm_{cv}}{dt} + \sum_{ex} \dot{m}_{ex} \tag{6.5}$$

where the subscripts *in* and *ex* denote inlet and exit, respectively, $\dot{m}$ represents the flow rate on a mass basis and *cv* denotes the control volume subsystem being considered. Next, the general energy balance equations applied to all system components is denoted as

$$\dot{Q} - \dot{W} + \sum_{in} \dot{m}_{in}\left(h_{in} + \frac{V_{in}^2}{2} + gZ_{in}\right) = \frac{dE_{cv}}{dt} + \sum_{ex} \dot{m}_{ex}\left(h_{ex} + \frac{V_{ex}^2}{2} + gZ_{ex}\right) \tag{6.6}$$

where the rate of heat input or output to the control volume per unit time is denoted with $\dot{Q}$, the input or output of work from the control volume is denoted by $\dot{W}$, the enthalpy is represented by *h*, the fluid velocity is represented by *V*, the elevation of the fluid is *Z*, and the gravitational constant is *g*. Moreover, the entropy balance is also applied to all system components that are written in its general form as

$$\sum_{in} \dot{m}_{in}s_{in} + \sum_{k} \frac{\dot{Q}_k}{T_k} + \dot{S}_{gen} = \frac{dS_{cv}}{dt} + \sum_{ex} \dot{m}_{ex}s_{ex} \tag{6.7}$$

where the specific entropy is represented by *s* and the entropy generation per unit timescale is denoted by $\dot{S}_{gen}$.

Each system component is also analyzed exergetically and the general exergy balance applied is denoted as

$$\dot{Ex}^Q + \sum_{in} \dot{m}_{in}ex_{in} = \sum_{ex} \dot{m}_{ex}ex_{ex} + \dot{Ex}_w + \dot{Ex}_{ds} + dEx_{cv}/dt \tag{6.8}$$

where the exergy associated with the heat transfer process $(\dot{Q})$ is represented by $\dot{Ex}^Q$, the specific exergy is represented by *ex*, the exergy associated with work input or output is denoted by $\dot{Ex}_w$, and the exergy destruction per unit time is written as $\dot{Ex}_{ds}$.

The system presented in Fig. 6.4 is analyzed dynamically considering the variations in the solar radiation intensities. The location of Ontario, Canada is chosen for the analyses and the radiation intensity variations on the average day of each month are considered, and the performance of the system is evaluated across the year accordingly. The overall performance of the system considering the operation on the considered day is evaluated energetically as

$$\eta_{ov} = \frac{\int \dot{W}_{MSTR}dt + \int \dot{W}_{FC}dt}{\int \dot{Q}_{s,in}dt + \int \dot{n}_{NH_3}\overline{LHV}_{NH_3}dt} \tag{6.9}$$

Also, the exergetic performance is determined according to the following equation:

$$\eta_{exov} = \frac{\int \dot{W}_{MSTR}dt + \int \dot{W}_{FC}dt}{\int \dot{Ex}^{\dot{Q}_{s,in}}dt + \int \dot{n}_{NH_3}\overline{ex}_{NH_3}dt} \tag{6.10}$$

where $\dot{W}_{MSTR}$ represents the power output of the MSRT, $\dot{W}_{FC}$ is the power output of the DAFC subsystem, $\dot{Ex}^{\dot{Q}_{s,in}}$ is the exergy associated with the solar thermal energy input, $\overline{LHV}_{NH_3}$ is the lower heating value on a per mole basis, $\dot{n}_{NH_3}$ denotes the molar input rate of ammonia fuel, $\overline{ex}_{NH_3}$ is the specific molar exergy of ammonia. The integral of these terms denotes the summation of these parameters throughout the day considering the dynamic variations in the solar radiation intensities.

The results of power outputs from the MSTR and DAFC subsystems evaluated on the average days of each month considering hourly variations are depicted in Fig. 6.5. Each graph depicts the results of 4 months showing the operation times as well as power output magnitudes of the MSTR as well as the DAFC. In the first 4 months of January–April, the power output from the MSTR is evaluated to reach a maximum value of nearly 130 MW in the month of February. Also, the least power output is evaluated to be nearly 80 MW. In addition to this, the operation times of the DAFC subsystem are associated with the sunset hours. When no solar radiation intensities are available, ammonia fuel is used through the DAFC subsystem to generate electrical power. The power output of the DAFC is set at 10 MW to maintain stable power output during the sunset hours from ammonia fuel. Furthermore, the power outputs in the next 4 months of May–August are depicted in Fig. 6.5B where the maximum power output is observed to be in the month of August where the peak power is slightly higher than in February. Further, the least MSTR power outputs are observed to occur in the months of April, May, and October. The power outputs from the turbine as well as the DAFC are used along with the corresponding incoming solar radiation intensities and input ammonia fuel required to determine the efficiencies that are depicted in Fig. 6.6. The energy efficiencies are depicted in Fig. 6.6A and the exergy efficiencies are presented in Fig. 6.6B. The results show that higher the solar radiation intensities, higher power outputs can be obtained. However, the efficiencies are lower for the summer months when high solar energy is available. This can be attributed to the lower utilization of solar thermal energy and conversion to electrical power during these months. Similarly, during the winter months of January and December, the efficiencies are observed to be higher than in other months of the year. This can also be attributed to the higher utilization of solar energy during these months. Also, the MSTR is operated for lower operation times in these

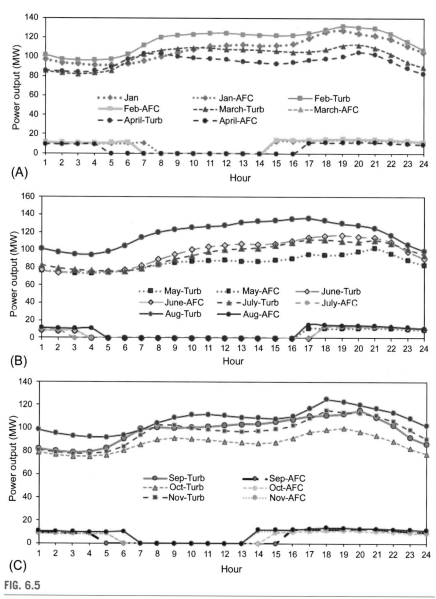

FIG. 6.5

Results of power outputs from the MSTR and DAFC subsystems as a function of time for the average days of each month.

months leading to lower exergy destruction. Although lower power outputs are obtained in these months, lower exergy destructions also take place that leads to higher utilization and thus efficiencies. The presented system thus provides a new technique to overcome the issue of intermittency in solar power plants. Solar radiation intensities vary across the year and entail the challenge of achieving

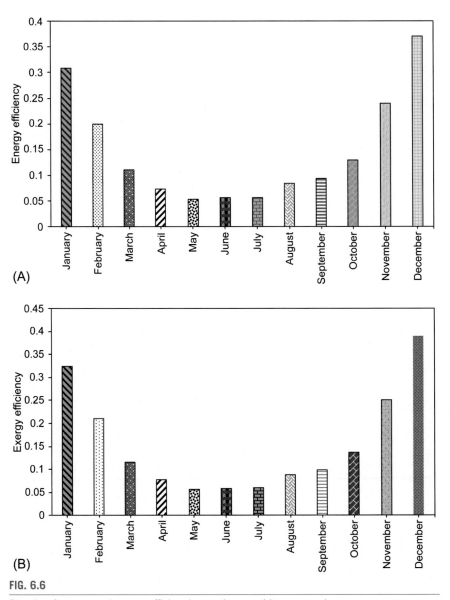

FIG. 6.6

Results of energy and exergy efficiencies on the monthly average days.

a stable and reliable power output supply. This is particularly important for the enhancement of renewable energy technologies across the globe. Solar energy comprises a clean and renewable method of electricity production and currently, solar photovoltaics are used more extensively than solar thermal power plants owing to the ease of using batteries for energy storage and demand-based electrical power supply. However, solar thermal technologies entail higher potentials of achieving higher overall efficiency as well as the usage of waste heat for producing useful commodities such as space heating, hot water, and cooling. Thus, the presented system can provide an effective way of enhancing the performance and operation of solar thermal power plants by mitigating the issue of intermittency through both TES as well as electrochemical AFC operation.

6.3 Solar tower-based multigeneration system integrated with DAFC

A solar tower-based integrated system designed for multigenerational purposes integrated with a direct feed-type AFC is depicted in Fig. 6.7. The solar tower power plant entails a hybrid molten salt TES and AFC system. Solar energy is focused on the solar tower via heliostat mirrors and this energy is used to heat the molten salt coming from the cold molten salt tank that entails a temperature of 400°C. During the day as well as during periods of high solar radiation availability, the molten salt

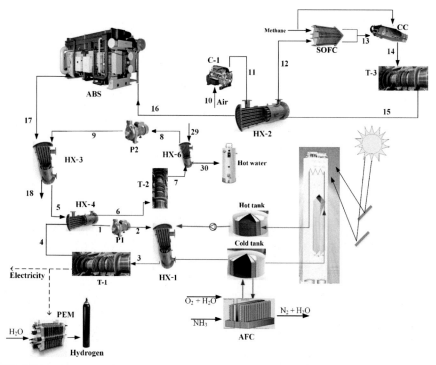

FIG. 6.7

Schematic representation of an integrated solar-based multigeneration system with molten salt AFC.

from this tank is sent to the solar tower and stored in a high-temperature hot molten salt tank. During periods of low or no solar availability, the molten salt-based AFC subsystem is operated to produce electrical power. This is integrated with the cold molten salt tank, which provides a flow of molten alkaline electrolyte comprised of 49% KOH and 51% NaOH by mole through the AFC. The electrolyte only provides a flow of ions and is not consumed during AFC operation. Furthermore, the hot molten salt is passed through heat exchanger HX-1 where the required energy is transferred to the steam Rankine cycle for power generation which operates between states 1 and 4. Water enters HX-1 at state 2 and is converted to superheated steam at state 3 before entering the steam turbine T-1, where electrical power is generated. The exit stream from T-1 leaves at state 4 before entering HX-4 where the waste heat is transferred to another Rankine cycle that operates between states 5 and 9. At state 9, water enters HX-3 where it absorbs the waste heat associated with the exhaust of the gas turbine T-3. Further, at state 5 the hot saturated water is passed through HX-4 where it converts into superheated vapor and enters the steam turbine T-2 at state 6. Moreover, the waste heat entailed in the exhaust of T-2 is utilized to produce hot water that is shown between states 29 and 30. At state 8, the saturated liquid water is pumped to a higher pressure and the secondary power generation cycle is repeated. In addition to this, the integrated system includes a solid oxide fuel cell and gas turbine (SOFC-GT) cycle that is depicted between states 10 and 16. At state 10, air enters the compressor C-1 that pressurizes it to the required operating pressure at state 11. Heat exchanger HX-2 is used to preheat the air before it enters the combined SOFC-GT subsystem. At state 12, the hot air enters the SOFC partially and the remaining is sent to the combustion chamber (CC). Also, both the SOFC as well as the CC are fed with methane gas as the fuel. After leaving the CC, the combustion gases enter the gas turbine T-3 where useful electrical power is generated. Once the exhaust gases leave T-3, they are passed through HX-2 to reheat the air before it enters the SOFC. Next, at state 16, the exhaust gases entail high thermal energy content that is used to operate an absorption cooling system (ABS), which intakes heat and provides cooling. Also, at state 17, the exhaust gases entail intermediate thermal energy content that is used to preheat water for the second Rankine cycle. Moreover, the system includes a proton exchange membrane (PEM) water electrolyzer for hydrogen production where a fraction of electrical power produced by T-1 is sent to PEM. Thus, the integrated multigeneration system with molten salt-based direct ammonia cell and SOFC-GT cycle provides useful commodities of electrical power, hydrogen, cooling effect, and hot water.

The system described above can be analyzed thermodynamically through both energy and exergy techniques. In these methods, energy balance, entropy balance, and exergy balances are applied to each system component to assess their thermodynamic performance. To apply these balances, several assumptions are made that represent actual operating conditions. The ambient conditions are taken as the environmental temperature of 25°C and ambient pressure of 100 kPa. Moreover, the pressure changing devices (turbines, compressors, throttle valves, and pumps) are analyzed under adiabatic conditions. However, the isentropic efficiency of mechanical devices (compressors, turbines, and pumps) is taken as 80%. Also, the heat exchangers are considered to entail no reductions in pressures of working fluids.

In addition to the above analyses, the PEM electrolyzer, SOFC, and AFC subsystems are also analyzed electrochemically similar to the electrochemical modeling discussed in the earlier chapters. The detailed analyses of the above subsystem can be obtained from our recent study [71]. The expression for the overall efficiency in terms of energetic assessment can be written as

$$\eta_{en,ov} = \frac{\dot{W}_{T1} + \dot{W}_{T2} + \dot{W}_{T3} + \dot{W}_{SOFC,AC} + \dot{W}_{AFC,AC} + \dot{Q}_{ABS} + \dot{m}_{H_2}LHV_{H_2} + \dot{m}_{29}(h_{30} - h_{29}) - \dot{W}_C}{\dot{Q}_{solar} + \dot{Q}_{CC} + (U)\dot{m}_{fu,SOFC}LHV_{CH_4} + \dot{m}_{NH_3}LHV_{NH_3}}$$

(6.11)

Also, the exergetic performance can be assessed in terms of the exergy efficiency as

$$\eta_{ex,ov} = \frac{\dot{W}_{T1} + \dot{W}_{T2} + \dot{W}_{T3} + \dot{W}_{SOFC,AC} + \dot{W}_{AFC,AC} + \dot{Q}_{ABS}\left(\frac{T_0}{T_{EV}} - 1\right) + \dot{m}_{H_2}ex_{H_2} + \dot{m}_{29}(ex_{30} - ex_{29}) - \dot{W}_C}{\dot{Q}_{solar}\left(1 - \frac{T_0}{T_{sun}}\right) + \dot{Q}_{CC}\left(1 - \frac{T_0}{T_{CC}}\right) + (U)\dot{m}_{fu,SOFC}ex_{CH_4} + \dot{m}_{NH_3}ex_{NH_3}}$$

(6.12)

where the power output from the turbine is denoted by $\dot{W}_T$, the power output from the SOFC after conversion to AC power is written as $\dot{W}_{SOFC,AC}$, the power output from the molten electrolyte AFC after conversion to AC power is written as $\dot{W}_{AFC,AC}$, the cooling provided through the ABS subsystem is denoted by $\dot{Q}_{ABS}$, the production rate of hydrogen on mass per unit time basis is written as $\dot{m}_{H_2}$, the lower heating value of hydrogen per unit mass is denoted as LHV_{H_2}, the power input supplied to the compressor is represented as $\dot{W}_C$, the total solar energy input to the integrated system is denoted as $\dot{Q}_{solar}$, the heat input to the CC is represented by $\dot{Q}_{CC}$, the fuel input to the SOFC is denoted by $\dot{m}_{fu,SOFC}$, the utilization factor is represented by U, the lower heating value and specific exergy of natural gas is denoted by LHV_{CH_4} and ex_{CH_4}, respectively, the mass input flow rate of ammonia to the AFC is denoted by $\dot{m}_{NH_3}$, and the lower heating value, as well as specific exergy of ammonia, are denoted by LHV_{NH_3} and ex_{NH_3}, respectively.

Moreover, the energetic performance of the ABS is determined as

$$COP_{ABS} = \frac{\dot{Q}_{ABS}}{\dot{Q}_{in,ABS}}$$

(6.13)

Similarly, the exergetic performance is analyzed as

$$COP_{ex,ABS} = \frac{\dot{Q}_{ABS}\left(\frac{T_0}{T_{EV}} - 1\right)}{\dot{Q}_{in,ABS}\left(1 - \frac{T_0}{T_{GEN}}\right)}$$

(6.14)

where COP denotes the coefficient of performance (COP), $\dot{Q}_{ABS}$ is the cooling provided by the ABS subsystem, $\dot{Q}_{in,ABS}$ is the required input of heat to the ABS system to generate the cooling effect, T_0 is the ambient temperature, T_{EV} is ABS evaporator temperature, and T_{GEN} is the temperature of the generator included in the absorption chiller. In addition, the energetic performance of the SOFC-GT subsystem is determined as

$$\eta_{SOFC/GT} = \frac{\dot{W}_{SOFC,AC} + \dot{W}_{T3} - \dot{W}_C}{\dot{Q}_{CC} + (U)\dot{m}_{fu,SOFC}LHV_{CH_4}} \qquad (6.15)$$

Also, the exergetic performance can be assessed as

$$\eta_{exSOFC/GT} = \frac{\dot{W}_{SOFC,AC} + \dot{W}_{T3} - \dot{W}_C}{\dot{m}_{fu,SOFC}ex_{CH_4}} \qquad (6.16)$$

The energetic and exergetic performance of the first Rankine cycle operating with turbine T-1 can be assessed as follows:

$$\eta_{FRC} = \frac{\dot{W}_{T1}}{\dot{Q}_{solar}} \qquad (6.17)$$

$$\eta_{exFRC} = \frac{\dot{W}_{T1}}{\dot{Q}_{solar}\left(1 - \dfrac{T_0}{T_{sun}}\right)} \qquad (6.18)$$

Also, the energetic performance in terms of efficiency for the second steam Rankine cycle operating with waste heat can be written as

$$\eta_{SRC} = \frac{\dot{W}_{T2}}{\dot{m}_9(h_5 - h_9) + \dot{m}_5(h_6 - h_5)} \qquad (6.19)$$

The exergetic performance is also expressed as follows:

$$\psi_{SRC} = \frac{\dot{W}_{T2}}{\dot{m}_9(h_5 - h_9)\left(1 - \dfrac{T_0}{T_{HX3}}\right) + \dot{m}_5(h_6 - h_5)\left(1 - \dfrac{T_0}{T_{HX4}}\right)} \qquad (6.20)$$

The energy efficiency of the integrated multigeneration system is evaluated through the above analyses as 39.1%. Also, the exergy efficiency is evaluated to be 38.7% for the integrated system under the operating conditions considered. Further information about the specific system parameters and operating conditions can be obtained from Ref. [71]. The efficiencies of the first Rankine cycle operating with solar energy input are found to be 19.9% and 21% evaluated energetically and exergetically, respectively. The SOFC-GT subsystem is found to be 55.9% and 68.5% efficient energetically and exergetically. Next, the performance of the ABS subsystem is also analyzed thermodynamically and 54%, as well as 31% energetic and exergetic coefficients of performances, are evaluated.

Several parametric analyses are also performed that provide the system performance results under different conditions of operation. The first parametric study is depicted in Fig. 6.8 where the effect of varying inlet pressure of T-1 on the overall system energetic and exergetic efficiencies is studied. In addition, the effects of this system parameters on other subsystems are also presented. The energetic performance of the integrated multigeneration system rises to 41.3% when the pressure at the inlet is raised to 3 MPa. Similarly, the exergetic performance rises to 41% in terms of efficiency when the inlet pressure is raised to this value. Also, the efficiencies of both steam Rankine cycle are observed to rise in rising inlet

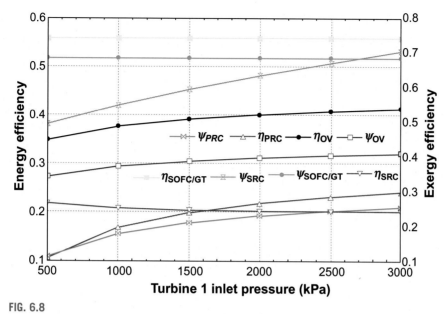

FIG. 6.8

Effect of varying turbine T-1 inlet pressure on the efficiencies.

pressure. This can be attributed to the higher power output of Turbine 1 at higher inlet pressures. Thus, the implementation of higher turbine inlet pressures can be implemented to achieve higher overall efficiencies for the developed system. The efficiency is observed to increase owing to the rise in power output of T-1 at higher inlet pressures. This is attributed to the change in the energy content of the working fluid at the turbine inlet. As the inlet pressure varies, the fluid energy content also changes that effects the power output of the turbine and thus the overall efficiencies. However, the energy efficiency of the secondary Rankine cycle is observed to follow a decreasing trend as the inlet pressure rises. This can be undertaken in the designed system as the overall efficiencies are rising.

The compressor C-1 entails a compression ratio that can be varied as an operating parameter by the usage of variable size compressors. This parameter effects the efficiencies of the overall system as well as subsystems as depicted in Fig. 6.9. The peak overall energetic and exergetic efficiencies are observed to be 39.2% and 38.8% that is associated with a compression ratio of 3. When the compression ratio rises higher than this value, the overall efficiencies are found to decrease. Nevertheless, the secondary Rankine cycle efficiencies rise with reducing the compression ratio. In addition, at a specific ratio of 2, the energetic and exergetic performances are observed to be 21% and 61%, respectively.

The system performance is also a function of the ambient temperature. As the ambient temperature varies, efficiencies are also effected. These are presented in Fig. 6.10, where the trends of changing efficiencies are depicted as a function of

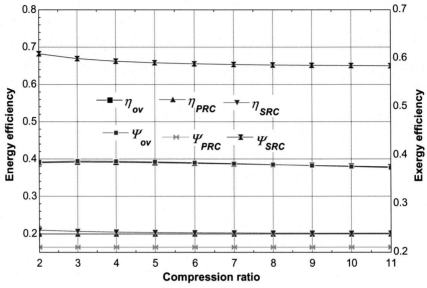

FIG. 6.9

Effect of changing compression ratio on the efficiencies.

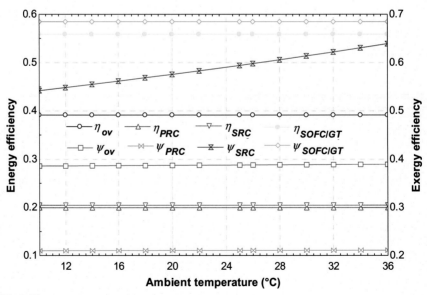

FIG. 6.10

Effect of ambient temperature on the efficiencies of the solar-based multigeneration system integrated with direct ammonia fuel cell.

the ambient temperature. Especially, the exergetic performance of the overall system rises as the ambient temperature elevates. For instance, when the temperature is 36°C, the exergetic performance is observed to be 38.9%. Similarly, the exergetic performance of the first Rankine cycle rises to 21%, the exergy efficiency of the secondary Rankine cycle increases to 63.9%, and the SOFC-GT exergetic efficiency increases to 68.5%. Further, at a lower ambient temperature of 10°C, the exergetic performance of the overall system reduces to 38.5%. Nevertheless, no considerable variation can be observed in energy efficiencies. This depicts the significance of implementing exergy analyses in the assessment of energy systems. As discussed earlier, exergy analysis also provides valuable information during the performance assessment of DAFCs. Especially, the exergy destruction rates in AFCs should be investigated under varying system operating conditions to analyze their overall exergetic performances.

6.4 Integrated ammonia internal combustion engine and DAFC-based cogeneration system

The schematic representation of the cogeneration system developed entailing integrated AFC and internal combustion engine (ICE) is depicted in Fig. 6.11. At state 1, ammonia from an ammonia tank enters the ammonia electrolysis cell (AEC) where the ammonia is dissociated into hydrogen and nitrogen molecules. The hydrogen molecules produced as a result of the dissociation process are sent to the ICE. The electrical input required to operate the AEC is obtained from a battery pack (BT). At state 25, ammonia is passed to the DAFC, which generates electrical power through a series of electrochemical reactions. However, the unreacted ammonia fuel at state 13 exits the FC and is stored in the aqueous ammonia storage (STR). This aqueous ammonia solution is utilized to obtain cooling. At state 14, the aqueous ammonia enters pump P1 where it is pressurized to a higher pressure before entering heat exchanger HX-1 where heat is transferred between the hot stream coming from the desorber (DES) and the cold stream coming from P1. After leaving HX-1 at state 16, the aqueous ammonia solution is sent to the desorber where water and ammonia molecules are separated through heat addition. The heat input required is obtained through the exhaust gases coming from the ICE that enter the DES at state 5 and exit at state 6 after transferring the required amount of thermal energy. The separated high concentration ammonia leaves DES at state 7 and the low concentration ammonia solution leaves at state 17 and enters HX-1 where it transfers heat to the incoming cold stream at state 15. After leaving HX-1, the pressure of the weak solution is dropped through the throttle valve (TV-1) before entering STR at state 19. Moreover, the strong ammonia solution at state 7 enters the regenerator where the remaining water molecules are separated from ammonia molecules and the water is recycled at state 8 to the DES. The ammonia is sent to the condenser (CON-1) where it condenses through heat rejection before entering the throttle valve (TV-2) where the pressure and thus the temperature is dropped before it enters the

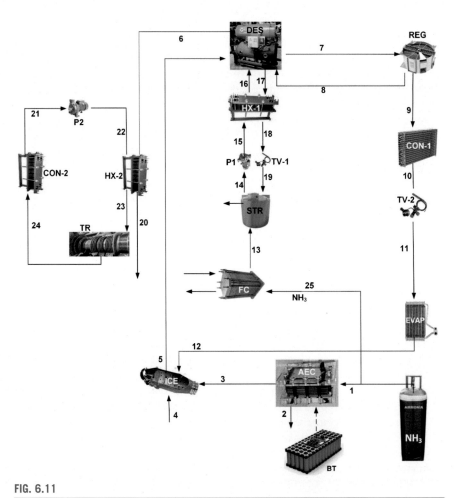

FIG. 6.11

Schematic representation of the integrated ammonia ICE and DAFC-based cogeneration system.

evaporator (EVAP) for providing cooling. Once the ammonia provides cooling through EVAP, it exits at state 12 and enters the ICE. The ICE subsystem entails inputs of hydrogen at state 3, ammonia at state 12, and air at state 4. Further, the exhaust gases resulting from the combustion reaction exit the ICE at state 5 at high temperatures. The high thermal energy content of stream 5 is utilized in the DES as well as HX-2. The waste heat recovered in the HX-2 is utilized for power generation through the steam turbine (TR). The steam Rankine cycle operates between states 21 and 24 to generate electrical power. Hence, the overall system outputs include power and cooling. The power is obtained through both the TR as well as ICE and the cooling is obtained in the EVAP.

The developed integrated system is analyzed thermodynamically through the balance equations listed earlier in Eqs. (6.1)–(6.4). Each system component is analyzed via thermodynamic energy and exergy techniques and further information about the comprehensive analyses, operating parameters utilized, and system modeling variables can be obtained from Ref. [72].

The energy efficiency of the hybrid ammonia and hydrogen fueled ICE is found as

$$\eta_{ICE} = \frac{\dot{W}_{ICE}}{\dot{N}_{A,in}\overline{LHV}_A + \dot{N}_{H,in}\overline{LHV}_H} \tag{6.21}$$

Also, the exergetic performance of the ICE can be determined as follows:

$$\eta_{ICE,ex} = \frac{\dot{W}_{ICE}}{\dot{N}_{A,in}\overline{ex}_A + \dot{N}_{H,in}\overline{ex}_H} \tag{6.22}$$

where the subsystem output is denoted by the numerator $\dot{W}_{ICE}$ that is the power output from the ICE, the input flow rates of ammonia and hydrogen are denoted by $\dot{N}_{A,in}$ and $\dot{N}_{H,in}$, respectively. The lower heating values per mole of ammonia and hydrogen are represented by $\overline{LHV}_A$ and $\overline{LHV}_H$. The specific molar exergy of these fuels is denoted by $\overline{ex}_A$ and $\overline{ex}_H$, respectively.

Further, the energy and exergy efficiencies of the power generation Rankine cycle are written as

$$\eta_{RC} = \frac{\dot{W}_{TR}}{\dot{Q}_{in,RC}} \tag{6.23}$$

$$\eta_{ex,RC} = \frac{\dot{W}_{TR}}{\dot{Q}_{in,RC}\left(1 - \dfrac{T_0}{T_{RC}}\right)} \tag{6.24}$$

where the power output of the steam turbine is represented with $\dot{W}_{TR}$, the thermal energy input to the power generation cycle is represented by $\dot{Q}_{in,RC}$, the reference temperature is denoted by T_0, and the average temperature at HX-2 is denoted by T_{RC}.

In addition to this, the energetic performance of the AEC subsystem is evaluated as

$$\eta_{AEC} = \frac{\dot{N}_H\overline{LHV}_H}{\dot{W}_{inAEC} + \dot{N}_A\overline{LHV}_A} \tag{6.25}$$

Similarly, the exergetic performance is also evaluated as follows:

$$\eta_{exAEC} = \frac{\dot{N}_H\overline{ex}_H}{\dot{W}_{inAEC} + \dot{N}_A\overline{ex}_A} \tag{6.26}$$

where the molar hydrogen production rate of the AEC is denoted as $\dot{N}_H$, the specific molar exergy and lower heating value are denoted by $\overline{ex}_A$ and $\overline{LHV}_A$, respectively, the molar input flow rate of ammonia is represented by $\dot{N}_A$, and the power input to the AEC is denoted by $\dot{W}_{inAEC}$.

Moreover, the energetic performance of the DAFC system is evaluated in terms of the electrical FC power output ($\dot{W}_{FC}$), input ammonia molar flow rate ($\dot{N}_{A,FC}$), and lower heating value ($\overline{LHV_A}$) as

$$\eta_{FC} = \frac{\dot{W}_{FC}}{\dot{N}_{A,FC}\overline{LHV_A}} \tag{6.27}$$

Similarly, the exergetic performance is evaluated as

$$\eta_{FC,ex} = \frac{\dot{W}_{FC}}{\dot{N}_{A,FC}\overline{ex_A}} \tag{6.28}$$

The cogeneration system entails the production of both electrical power and cooling. Hence, it is essential to determine the energetic and exergetic performance of the cooling subsystem. The COP for the energetic performance of the cooling subsystem is written as

$$COP_{CS,en} = \frac{\dot{Q}_{CS,EV}}{\dot{Q}_{des}} \tag{6.29}$$

where the cooling effect provided per unit time in the evaporator is denoted by $\dot{Q}_{CS,EV}$ and the corresponding heat input to the cooling cycle at the desorber is represented by $\dot{Q}_{des}$. Similarly, the exergetic COP is determined as

$$COP_{CS,ex} = \frac{\dot{Q}_{CS,EV}\left(\frac{T_0}{T_{EV}} - 1\right)}{\dot{Q}_{des}\left(1 - \frac{T_0}{T_{des}}\right)} \tag{6.30}$$

where the temperature of the evaporator is written as T_{EV} and the temperature of the desorber is written as T_{des}.

It is important to analyze the performance of the overall cogeneration system in terms of the energetic and exergetic evaluations. The overall performance in terms of energy efficiency is written as

$$\eta_{ov,co} = \frac{\dot{W}_{TR} + \dot{W}_{FC} + \dot{W}_{ICE} + \dot{Q}_{CS,EV}}{\dot{N}_A\overline{LHV_A} + \dot{W}_{AEC}} \tag{6.31}$$

Also, the overall exergy efficiency is also determined as follows:

$$\eta_{ov,ex,co} = \frac{\dot{W}_{TR} + \dot{W}_{FC} + \dot{W}_{ICE} + \dot{Q}_{CS,EV}\left(\frac{T_0}{T_{EV}} - 1\right)}{\dot{N}_A\overline{ex_A} + \dot{W}_{AEC}} \tag{6.32}$$

where the useful outputs of the system are denoted in the numerator including the turbine power output ($\dot{W}_{TR}$), fuel cell power output ($\dot{W}_{FC}$), ICE power output ($\dot{W}_{ICE}$), and the cooling effect provided ($\dot{Q}_{CS,EV}$). The system inputs needed to achieve these useful outputs include the product of molar ammonia input flow rate ($\dot{N}_A$) and lower heating value ($\overline{LHV_A}$) for energy efficiency, and specific molar exergy ($\overline{ex_A}$) for exergy efficiency. In addition, the external power input required for the AEC ($\dot{W}_{AEC}$) is also considered in both energetic as well as exergetic performance evaluation.

The system modeling results provide an overall efficiency value of 59.9% determined energetically and 51.9% analyzed exergetically. Further, the energetic performance of the power generation Rankine cycle (η_{RC}) is determined as 25.5%, whereas the exergetic performance of this cycle ($\eta_{ex,RC}$) is 38.2%. Moreover, the AEC is found to entail high performances both energetically and exergetically, where the AEC energy efficiency (η_{AEC}) is evaluated as 93.9% and the exergetic performance is determined as 90.9%. In addition to this, the DAFC is also evaluated energetically and exergetically. The energy efficiency of the FC subsystem is 44.4%, whereas the exergetic performance is evaluated as 41.7% Moreover, the exergetic performance of the ICE subsystem fueled with hydrogen and ammonia is determined to be 43.8%. The exergy destruction rate in all system components is also evaluated. The highest rate of exergy destruction is found in the ICE subsystem where 1308 kW of exergy is destroyed per unit time. Also, HX-2 is found to entail high exergy destruction rate of 997.9 kW. This can be attributed to the high difference in temperatures between hot and cold streams and can thus be minimized by decreasing the temperature differences. The AEC and TR subsystems are found to be associated with exergy destruction rates of 136.1 and 377.5 kW, respectively. The exergy destruction in the AEC can be reduced by decreasing the polarization losses, whereas the exergy destroyed in the turbine can be decreased by developing turbines with higher isentropic efficiencies. Higher isentropic efficiency signifies lower irreversibilities and thus lower exergy destruction rates. The thermophysical properties of each state point are provided in Ref. [72]. The steam turbine is evaluated to provide a power output of 1848 kW and power output of the ICE is determined to be 11,145 kW. Further, the fuel cell is evaluated to entail an electrical power output of 19.7 kW and the cooling effect provided at the evaporator is 1290 kW. The AEC power consumption is determined as 1.85 kW. The ICE operating pressure is set at 250 kPa and fuel blend ratio is considered to be 1% hydrogen by mass. Also, the air-fuel ratios of ammonia and hydrogen are 6.06 and 34.2, respectively. For the AEC subsystem, the inlet pressure of ammonia fuel is set at 870 kPa and the operating temperature is considered as 25°C. The exchange current density of the AEC is taken as 0.37 A/m^2, whereas the operating current density is considered to be 2500 A/m^2. These system parameter values have been derived from the literature and are further described in the reference above. The system performance is also analyzed at varying operating conditions. Some of these parametric studies are depicted in Figs. 6.12–6.14. The effect of varying ambient temperature on the overall efficiencies as well as the efficiencies of the power generation cycle are depicted in Fig. 6.12. The ambient temperature is varied from 0 to 40°C and both energetic as well as exergetic performances are studied. The overall exergy efficiency increases continuously with rising ambient temperature. For example, at 0°C the overall exergy efficiency is nearly 51% and when a higher temperature of 40°C exists, the exergy efficiency of the overall system

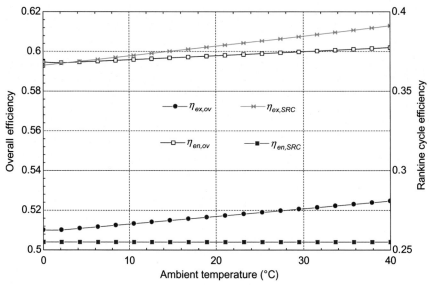

FIG. 6.12

Variation of overall system and Rankine power generation cycle efficiencies with ambient temperature.

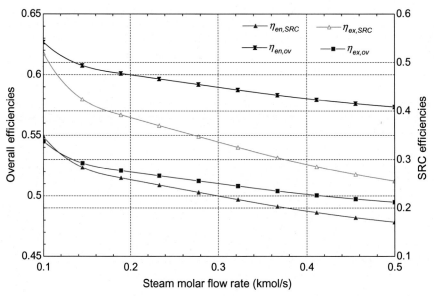

FIG. 6.13

Variation of overall system and Rankine power generation cycle efficiencies with steam molar flow rate.

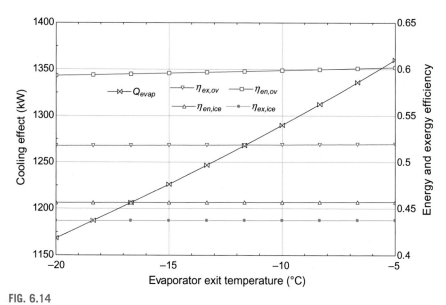

FIG. 6.14

Variation of cooling effect, overall efficiencies and internal combustion engine efficiencies with change in evaporator exit temperature.

rises to 53%. The energetic performance, however, does not show the same difference in efficiencies with temperature variations. The energy efficiency at 0°C is evaluated to be 59.4%, which is found to rise to 60.2% at an elevated ambient temperature of 40°C. Also, the exergetic performance of the power generation Rankine cycle is enhanced as the ambient temperature elevates. The exergetic efficiency, for instance, is evaluated as 36.6% at 0°C. Nevertheless, as the ambient temperature is raised to 40°C, the exergy efficiency attains a higher value of 39.1%. The energy efficiency of the power generation Rankine cycle, on the other hand, shows little variation with changing ambient conditions. This depicts the importance of exergy analyses in the thermodynamic assessment of energy systems. The specific exergy at each state point varies as the ambient conditions change, thus neglecting exergy analyses neglect the assessment of irreversibilities occurring in system components.

In addition, the flow rate of steam or water in the power generation Rankine cycle effects the performance of the subsystem as well as overall system. This is depicted in Fig. 6.13, where the effect of steam molar flow rate on the energetic as well as exergetic efficiencies is studied. A decreasing trend is observed in the efficiencies with rising molar flow rates. At a molar flow rate of 0.1 kmol/s, the energetic performance of the overall system is evaluated to be 62.7% in terms of efficiency. This is found to decrease by 5.4% when a higher molar flow rate of 0.5 kmol/s is considered. Similarly, the exergy efficiency at 0.1 kmol/s of steam flow rate is evaluated as 54.5% and this decreases when the flow rate is raised. An efficiency of 49.5% is evaluated exergetically for the overall system when a steam flow rate of 0.5 kmol/s is considered. In addition to this, the efficiencies of the power generation cycle are also observed to reduce with rising flow rate.

At 0.5 kmol/s of steam flow rate, the energetic performance of the power generation cycle is evaluated as 17.1% that is seen to increase considerably when a lower molar flow rate of 0.1 kmol/s is utilized where the energetic efficiency rises to 34.7%. Similarly, the exergetic efficiency varies between 25.6% and 51.9% for the same variation in molar flow rates.

The temperature chosen for the evaporator operation that provides the cooling effect also effects the performance of the overall system as well as other system components. This is depicted in Fig. 6.14 where the evaporator exit temperature is varied between −20 and −5°C and the effects on the amount of cooling effect provided, overall energetic as well as exergetic efficiency are studied. Also, the performance of the ICE is analyzed in terms of efficiencies for the same variation in evaporator exit temperature. The cooling effect provided in the evaporator is directly related to the exit temperature at the evaporator. For instance, when the exit temperature is raised from −20 to −5°C, the amount of cooling provided rises from 1168.2 to 1360.2 kW. In addition to this, energetic performance of the ICE is observed to marginally drop with a decrease in evaporator temperature. The energy efficiency of the ICE, for instance, reduces by 0.1% when the evaporator exit temperature decreases by 15°C. This can be attributed to the integrated configuration of the evaporator and ICE as can be depicted from Fig. 6.5 where the exit stream of the evaporator at state 12 directly enters the ICE. The exergetic performance of the ICE is also observed to show similar variation in efficiency with the same change in evaporator temperature. The evaporator exit temperature defines the ICE inlet ammonia temperature. Hence, the temperature variations lead to changes in the specific exergy of the ammonia stream that effect the performance of the subsystem. As far as the overall system is concerned, the energetic performance is observed to be an efficiency value of 60.2% at −5°C, which is found to reduce by 0.9% when the evaporator temperature is decreased to −20°C. Similarly, the exergetic efficiency of the overall system is evaluated to decrease by 0.1% for the same change in the evaporator exit temperature.

6.5 Solar-wind-based integrated system utilizing AFCs for energy storage to produce electricity and freshwater

The integrated renewable energy system is developed based on solar and wind energy resources integrated together to produce fresh water and electricity, utilizing AFCs for energy storage is depicted in Fig. 6.15. The heliostat mirror field (HF) reflects the incoming solar radiation onto a central solar tower receiver (ST) that entails a circulation of molten salt heat transfer medium that absorbs the incoming energy and transfers it to the MSRT steam Rankine cycle (MSTR). The MSTR entails the reheat Rankine cycle that reheats the steam twice after exiting each pressure level turbine. At state 3, superheated water vapor enters the high-pressure stage

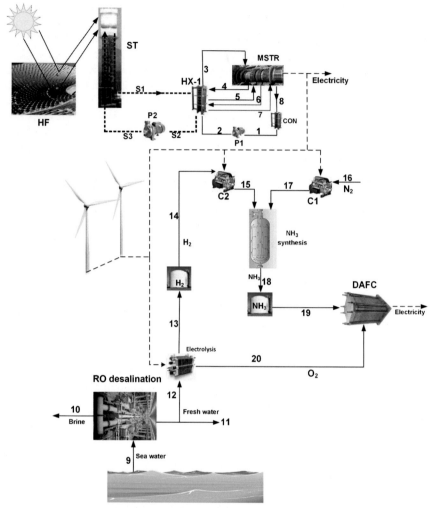

FIG. 6.15

Schematic representation of the renewable energy-based system integrated with AFCs for electricity and freshwater production.

of the turbine and exits at state 4 where it is passed through HX-1 to be reheated to state 5 before entering the medium pressure stage of the MSTR. Further, at state 6, the steam is again reheated before entering the low-pressure level of MSTR at state 7. At state 8, saturated steam exits the MSTR and is passed through the condenser (CON) for heat rejection before entering pump P1, which is used to attain the high pressures needed for the operation of the power generation cycle. The electrical power output from the MSTR is used as a useful commodity and when excess solar energy is available, electricity is passed unto the ammonia synthesis and storage system. In addition to solar-based power generation, wind turbines are used to generate electrical power. The electricity output from the system comprises both wind and solar energy-based power generation and during excess solar or wind energy, the excess energy is utilized to synthesize ammonia. The ammonia synthesized is stored for later usage when low wind or solar energy would be available. Moreover, the renewable energy-based system also entails the production of fresh water via reverse osmosis (RO) desalination where saline water enters the RO plant at state 9 and through a series of treatment processes, fresh water exits at states 11 and 12. A portion of the fresh water produced from the RO plant is sent to the PEM water electrolysis (WES). The PEM electrolysis subsystem dissociates water molecules in hydrogen and oxygen.

The hydrogen exits the subsystem at state 13 and the oxygen exits at state 20. When excess power is available either from the solar or wind power plants, it is sent to the compressors C1 and C2 as well as the RO desalination system. The compressor C2 pressurizes hydrogen gas to the required ammonia synthesis pressure before entering the synthesis reactor at state 15. Simultaneously, nitrogen gas is obtained from an external input and is pressurized to the required synthesis pressure by C1 before it enters the reactor at state 17. The synthesized ammonia is stored for later utilization. When solar intensities are low, during sunset hours, and when wind velocities are low the stored ammonia is sent to a DAFC entailing a solid oxide electrolyte. The DAFC operates at high operating temperatures and produces electrical power output via ammonia fuel. The electrochemical interactions occurring in the solid oxide electrolyte-based DAFC were discussed in earlier chapters. The system is modeled for varying solar radiation intensities as well as wind velocities and the system performance is analyzed across the year. The location of Toronto, Canada is chosen to analyze the developed system and the performances are assessed considering the solar intensities as well as wind velocities on the average days of each month. The hourly values of intensities and velocities on the average days are considered to perform a transient simulation of the system across the year. The variations in the production of electrical power, hydrogen and ammonia are first analyzed considering the hourly variations on the average days of each month. The performance of the developed system is then analyzed for each hour through thermodynamic assessments and the monthly energetic and exergetic performances in terms of efficiencies are analyzed. The thermodynamic balance equations described earlier are used and further information can be obtained from Ref. [73].

The efficiencies of the subsystems, as well as overall system, are determined considering the annual analysis. The energetic performance of the MSTR is determined as

$$\eta_{MSTR} = \frac{\int \dot{W}_{o,MSTR} dt}{\int \dot{Q}_{in,MSTR} dt} \tag{6.33}$$

where $\int \dot{W}_{o,MSTR} dt$ denotes the integral of the power output from the MSTR that is a function of varying solar intensities and $\int \dot{Q}_{in,MSTR} dt$ represents the integral of the heat input to the MSTR cycle that is also a function of the intensities. Similarly, the exergetic MSTR performance is evaluated as

$$\eta_{ex,MSTR} = \frac{\int \dot{W}_{o,MSTR} dt}{\int \dot{Q}_{in,MSTR} \left(1 - \frac{T_0}{T_{HX1}}\right) dt} \tag{6.34}$$

where T_0 is the ambient temperature and T_{HX1} denotes the average temperature in HX-1. Moreover, the heat input to the MSTR can be evaluated as

$$\int \dot{Q}_{in,MSTR} \, dt = \int \dot{m}_{S-1} h_{S-1} dt - \int \dot{m}_{S-2} h_{S-2} dt \tag{6.35}$$

where h denotes the specific enthalpy, $\dot{m}$ represents the mass flow rate and the subscripts represent the state points.

Further, the wind turbine farm also entails a performance ratio in terms of the wind energy input and electrical power output. This is evaluated as

$$\eta_{WT} = \frac{\int \dot{W}_{o,WT} dt}{\int 0.5\rho A V^3 N_{WT} dt} \tag{6.36}$$

where $\dot{W}_{o,WT}$ denotes the power output from the wind turbine farm, ρ represents the density of air, V is the wind velocity, A is the cross-sectional area of turbine blades, and N_{WT} represents the total number of wind turbines considered.

The RO water treatment also entails an important subsystem that provides fresh water. The performance of this subsystem is analyzed energetically as

$$\eta_{FW} = \frac{\int \dot{m}_{11} h_{11} dt}{\int \dot{W}_{in,FW} + \int \dot{m}_9 h_9 dt} \tag{6.37}$$

where the freshwater output from the overall system is denoted by $\dot{m}_{11}$, the specific enthalpy of this fresh water is represented by h_{11}, the total power input to the RO subsystem is represented by $\dot{W}_{in,FW}$, the saline water input and its enthalpy is denoted by $\dot{m}_9$, and h_9, respectively. The exergetic performance of the system is also assessed as

$$\eta_{exFW} = \frac{\int \dot{m}_{11} ex_{11} dt}{\int \dot{W}_{in,FW} + \int \dot{m}_9 ex_9 dt} \tag{6.38}$$

In addition to this, the exergetic performance of the ammonia synthesis subsystem is determined according to

$$\eta_{exAS} = \frac{\int \dot{m}_{18} ex_{18} dt}{\int \dot{m}_{15} ex_{15} dt + \int \dot{m}_{17} ex_{17} dt} \tag{6.39}$$

where the ammonia output from the subsystem is denoted by $\dot{m}_{18}$, the specific exergy at this state point is represented by ex_{18}, the hydrogen input is represented by $\dot{m}_{15}$, the specific exergy of hydrogen is written as ex_{15}, the nitrogen input mass flow rate is written as $\dot{m}_{17}$, and the corresponding specific exergy is denoted as ex_{17}.

Furthermore, the energetic performance of the DAFC is evaluated in terms of the electrical power output ($\dot{W}_{DAFC}$), ammonia molar input flow rate ($\dot{N}_{18}$), and its lower heating value ($\overline{LHV}_{NH_3}$) as

$$\eta_{DAFC} = \frac{\int \dot{W}_{DAFC}\,dt}{\left(\int \dot{N}_{18}dt\right)\left(\overline{LHV}_{NH_3}\right)} \qquad (6.40)$$

Similarly, the exergetic performance is also evaluated as follows:

$$\eta_{exDAFC} = \frac{\int \dot{W}_{DAFC}\,dt}{\left(\int \dot{N}_{18}dt\right)\left(\overline{ex}_{NH_3}\right)} \qquad (6.41)$$

where $\overline{ex}_{NH_3}$ denotes the specific molar exergy of ammonia. The overall system performance is important to be analyzed both energetically and exergetically. The overall energy efficiency is determined from

$$\eta_{sys,ov} = \frac{\int \dot{W}_{GR,MSTR}dt + \int \dot{W}_{GR,WF}dt + \int \dot{W}_{DAFC}\,dt + \int \dot{m}_{11}h_{11}dt}{\int \dot{Q}_{in,sol}dt + \int 0.5\rho AV^3 N_{WF}dt + \int \dot{m}_9h_9dt} \qquad (6.42)$$

where $\dot{W}_{GR,MSTR}$ represents the power output of the MSTR subsystem sent to the grid, $\dot{W}_{GR,WT}$ denotes the power output from the wind turbine farm utilized for grid electricity supply, $\dot{W}_{DAFC}$ is the power output from the DAFC, $\dot{m}_{11}h_{11}$ is the energy content of the fresh water produced from the RO subsystem, $\dot{Q}_{in,sol}$ is the solar energy input to the system, N_{WF} is total number of wind turbines in the farm, and $\dot{m}_9h_9$ denotes the energy content of the saline water entering the system.

The exergetic performance of the overall system is also assessed by using

$$\eta_{sys,ex,ov} = \frac{\int \dot{W}_{GR,MSTR}dt + \int \dot{W}_{GR,WF}dt + \int \dot{W}_{DAFC}\,dt + \int \dot{m}_{11}ex_{11}dt}{\int \dot{Q}_{in,sol}\left(1-\frac{T_0}{T_{sun}}\right)dt + \int 0.5\rho AV^3 N_{WF}dt + \int \dot{m}_9ex_9dt} \qquad (6.43)$$

where T_0 is the ambient temperature, T_{sun} is the sun temperature, ex_{11} is the specific exergy at state 11, and ex_9 is the specific exergy at state 9. Hence, in this way the performances are assessed by taking the summation of the useful outputs produced and the required inputs utilized for each day considering the variations in the solar intensities and wind velocities. The algorithm implemented for system modeling is depicted in Fig. 6.16. The threshold of excess power generation is set at 10 MW. When the combined solar and wind power output exceeds this limit, the remaining

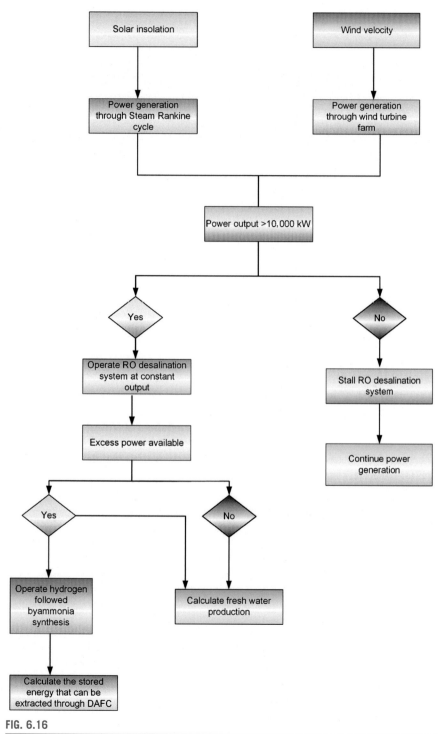

FIG. 6.16

Flowchart depicting the implemented algorithm for system modeling.

power is directed toward the RO desalination plant to produce fresh water with a specified output capacity. Once the set output is attained, the remaining excess power is sent to the ammonia synthesis subsystem. This subsystem includes the PEM water electrolyzer as well as compressors C1 and C2. The PEM electrolyzer produces hydrogen that is stored in a tank and is passed to C2 to pressurize it to the required ammonia synthesis pressure. Also, C1 pressurizes nitrogen gas to the appropriate pressure before it enters the synthesis reactor, where nitrogen is considered as an external input to the system. The ammonia synthesized during these excess power periods is used to evaluate the potential of power generation through the DAFC. When the total power generation from the solar and wind resources is lower than the threshold of 10 MW, the electricity produced is sent to the grid and electrical power is the only useful output.

The results of the total power output from the integrated solar and wind energy resources are depicted in Fig. 6.17. These have been evaluated considering actual wind velocities and modeled solar radiation intensities for the location of Toronto, Canada. As can be observed from the figure, the summer month of June provides a considerably higher power output along with the month of August. These are attributed to the high wind velocities as well as solar radiation intensities during these months. The maximum power output from the hybrid plant is found to reach 62 MW in June and nearly 65 MW in August. The integrated system is designed in a way that the winter months also exceed the threshold of 10 MW power output for a few hours during the day depending on the solar radiation intensities. For instance, in the winter month of January, the threshold is exceeded during the day when solar intensities are available and during this time, fresh water and ammonia are synthesized. Moreover, results of ammonia synthesis rates for monthly average days is depicted in Fig. 6.18 where the results are proportional to the excess power availability. The maximum synthesis rate of ammonia is observed to occur in the months of June and August where the production rates of nearly 50 mol/s are achieved. These provide an encouraging system result that the developed system can be effectively implemented in hybrid solar and wind power plants. Moreover, the maximum ammonia production rate during the winter month of November where low solar intensities, as well as wind velocities occurred, was nearly 5 mol/s. Depending on the availability of solar and wind energy, ammonia production rates vary in the present system unlike conventional ammonia synthesis plants that entail fixed capacities for ammonia synthesis. However, renewable energy-based ammonia synthesis systems will entail this challenge of intermittency and can be addressed in several ways. Similar to conventional solar photovoltaic power plants where battery storage is used to address the issue of intermittency, DAFCs can be used in the present system to overcome such issues. Also, additional ammonia input from external sources can be provided in case there is considerably low availability of both solar and wind energy. Nevertheless, in the present system designed, each monthly average day entails excess energy that is used to synthesize ammonia and use it for power generation later.

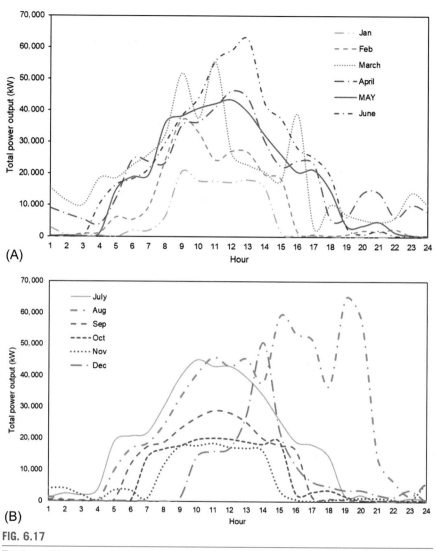

FIG. 6.17

Total monthly power outputs results on the average days for (A) January–June and (B) July–December.

The overall system performance results are summarized in Fig. 6.19 in terms of the energetic and exergetic efficiencies. The energy efficiency value is observed to entail a value of 48.1% at the least point that occurs in the month of June. Further, the maximum efficiency point is evaluated as 53.3% that is observed to occur in the month of March. Although in June, the highest total power output was evaluated along with the highest ammonia synthesis rate, the efficiency entails a lower value

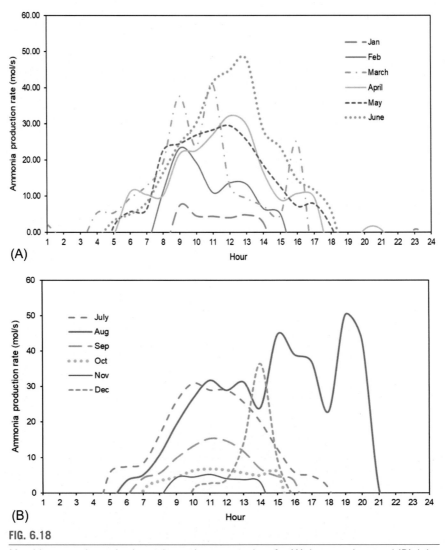

FIG. 6.18

Monthly ammonia synthesis results on the average days for (A) January–June and (B) July–December.

owing to the comparatively higher input of energy due to high wind velocities and solar intensities. Thus, although higher ammonia is synthesized and thus higher energy is obtained from the DAFC, higher energy losses and exergy destructions due to an increase in the operation time of each system component result in comparatively lower energy efficiency. On the other hand, although the winter months of January and December entail lower ammonia production, they entail higher efficiencies than the month of June. This is attributed to the lower amount of exergy

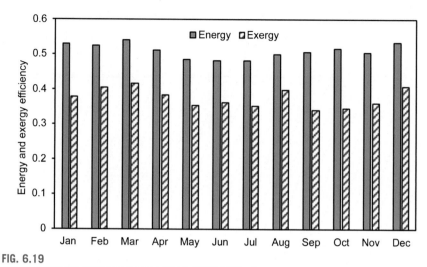

FIG. 6.19

Efficiency results for the monthly average days of (A) January–June and (B) July–December.

destructions as well as energy losses due to lower operation time of each system component. The actual operation of each system component is associated with the entropy generation and thus exergy destruction. Hence, in the winter months, lower outputs resulting from lower inputs lead to higher performance ratios of useful output and required input leading to higher efficiencies. Thus, the developed system provides an innovative methodology to utilize solar and wind energy resources integrated with DAFCs.

6.6 Integrated ammonia synthesis and DAFC system based on solar photovoltaic energy

The developed integrated ammonia synthesis and fuel cell system utilizing solar photovoltaic (PV) energy is depicted in Fig. 6.20. The system entails the synthesis of ammonia during periods of excess solar energy and thus excess solar PV power. The electricity from the PV power plant is utilized primarily for electrical power output. However, during excess solar energy availability, the generated electrical power is sent to the ammonia synthesis system where ammonia is synthesized and stored for later usage. When electrical power is needed during periods of none or low availability of solar energy, the stored ammonia is utilized in a direct type AFC to generate electricity. At state 1, water enters the PEM water electrolyzer which intakes electricity during excess solar availability and produces hydrogen gas (state 2) that is stored in hydrogen tanks. Further, the pressure swing adsorption (PSA) subsystem intakes air at state 4 along with electrical input to generate a stream of nitrogen at

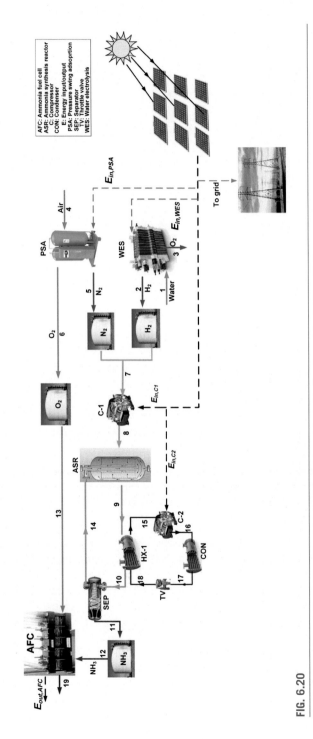

FIG. 6.20

Schematic representation of the developed integrated ammonia synthesis and fuel cell system.

state 5 as well as oxygen at state 6. The nitrogen gas produced is stored in the nitrogen storage tank and the oxygen produced is also stored. Moreover, the hydrogen and nitrogen gases are passed through compressor C-1 at state 7 to pressurize the mixture to the appropriate pressure needed for ammonia synthesis. At state 8, the nitrogen and hydrogen mixture enters the ammonia synthesis reactor (ASR) and reacts chemically to produce ammonia. The ammonia produced along with the unreacted hydrogen and nitrogen exits the ASR at state 9 and are passed through heat exchanger HX-1 where the produced ammonia is condensed and the remaining unreacted gases are sent back to the ASR for further usage to synthesize ammonia. In this way, during excess solar energy availability, electrical power output of the solar PV is sent to the hydrogen production via WES, nitrogen generation via PSA, and gaseous compression via C-1 subsystems. The synthesized ammonia is separated in the separator (SEP) at state 11 and is stored for later usage. At state 12, the stored ammonia is sent to the DAFC for power generation.

When low solar availability occurs, the stored ammonia is used to generate clean and renewable power. The oxidant required for fuel cell operation enters at state 13 and is obtained from the PSA subsystem that separates air into nitrogen and oxygen. At state 6, the oxygen produced in the PSA subsystem is stored in the oxygen storage tank. The stored oxygen inputs the DAFC subsystem at state 13. The ammonia separation process also requires a power input where the refrigeration cycle based on the R-134a refrigerant operates between states 15 and 18. At state 18, cold refrigerant enters HX-1 that cools the gaseous mixture coming from the ASR at state 9. The refrigerant exiting HX-1 is compressed to a higher pressure before it is sent to the condenser (CON) where it rejects heat before entering the throttle valve (TV) at state 17. The TV throttles the refrigerant to a lower pressure and thus lower temperature at state 18 before it enters HX-1 to further cool the gaseous mixture incoming from the ASR.

The developed integrated ammonia synthesis and fuel cell system is also analyzed dynamically considering the variations in solar radiation intensities across the year. Monthly average days of each month are taken to perform the dynamic simulation of the system. The city of Toronto, Canada is chosen for the modeling and the simulation is conducted by evaluating the solar radiation intensities at the location at each hour as described earlier in Section 6.3. Each system component is analyzed thermodynamically similar to the previous systems discussed. The overall system performance is assessed in terms of the energetic as well as exergetic efficiencies for each monthly average day. The equation for the assessment of the overall performance in terms of energy efficiency is written as

$$\eta_{ov} = \frac{\int_{t=1}^{n_{el}} \dot{P}_{PV,el} dt + \int_{t=1}^{n_{AFC}} \dot{P}_{NH_3,AFC} dt}{\int_{t=1}^{n_s} \dot{Q}_s dt} \tag{6.44}$$

where the electrical power output from the PV system utilized as electricity directly is written as $\dot{P}_{PV,el}$, the electrical power output obtained from the DAFC is denoted as $\dot{P}_{NH_3,AFC}$, and the solar radiation intensity at a given hour is represented as $\dot{Q}_s$. The exergetic performance is also assessed that is written as

$$\eta_{ex,ov} = \frac{\int_{t=1}^{n_{gd}} \dot{P}_{PV,gd}dt + \int_{t=1}^{n_{AFC}} \dot{P}_{NH_3,AFC}dt}{\int_{t=1}^{n_s} \dot{Q}_s\left(1 - \frac{T_0}{T_{sn}}\right)dt} \tag{6.45}$$

where the exergy input to the system is placed in the denominator that is evaluated as shown in the equation from the thermal energy input rate ($\dot{Q}_s$), temperature of the sun (T_{sn}), and the ambient temperature (T_0). The ambient temperature for the chosen location of Toronto is obtained for each month considered to evaluate the exergetic performance. The above equations describe the overall system performance considering the energy input during the charging phase and the energy output during the discharging phase as well as under normal PV operation.

The simulation results of power output from the PV plant are depicted in Fig. 6.21 where the power outputs on the monthly average days are evaluated considering the hourly solar radiation intensity variations. The power output at the maximum level is evaluated during the month of June, where the output of 346.2 MW is observed in the month of June. Also, the maximum output of power during the winter month of December is evaluated to be 155.3 MW. This month entailed the lowest solar

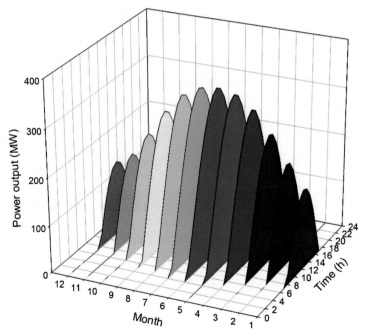

FIG. 6.21

Results of power output from the solar PV plant on monthly average days.

radiation intensity potential as compared to other months of the year as can be observed in the figure. Further, the number of daylight hours are also decreased for the winter months where the daylight hours are observed to be 8. However, the summer months are associated with higher daylight hours. For instance, the month of June is associated with 16 daylight hours. These factors are important for the design of the integrated ammonia synthesis and fuel cell system. The components of the overall system need to be sized accordingly that can incorporate the power input provided. To model the developed system, the cutoff threshold of 90% of the maximum power output is applied. Under this criteria, when the electrical power output from the PV plant exceeds 90% of the peak power value, the electrical supply to the integrated ammonia synthesis and fuel cell system is started and ammonia is synthesized. The produced ammonia is stored in an ammonia storage tank until low solar availability occurs when ammonia fuel is input to the DAFC subsystem to generate electrical power.

The results of ammonia synthesis during the hours with more than 90% of the peak power are depicted in Fig. 6.22. The summer months entail higher ammonia synthesis rates as well as total production amounts owing to higher solar availability. The rate of ammonia synthesis at the maximum point across the year reaches a value of 64.8 mol/s that occurs in June. Further, the synthesis rate in the month with low solar intensities is associated with lower synthesis rates proportionately. For

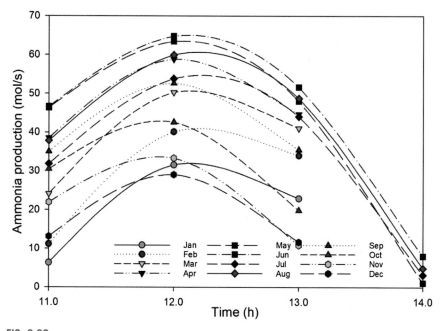

FIG. 6.22

Results of ammonia production rates during periods of excess PV power.

instance, in January the maximum rate of ammonia synthesis reaches a value of 31.6 mol/s. In addition to this, the synthesis rates vary according to the amount of excess electrical power available. However, the total ammonia synthesized during a day depends on both the synthesis rates during the excess hours as well as the number of hours of available excess power. The number of excess hours in summer months is observed to be nearly 4 h whereas during the winter months these are found to decrease to 3 h. Therefore, the developed system can also be implemented in applications where ammonia can be utilized as a useful commodity. For instance, in fertilizer production plants where significant amounts of ammonia need to be produced, the developed system can be implemented where the clean synthesis of ammonia occurs through the utilization of excess power. The electrical power needs of the fertilizer plant can be met through the solar PV plant and the excess power can be used to produce ammonia. This is being proposed owing to the considerable environmental emissions associated with the current conventional method of producing ammonia, which relies heavily on fossil fuels. Through the implementation of the present system, a portion of the ammonia produced by the fertilizer plants can be made environmentally benign. Also, depending on the amounts synthesized, a portion can also be utilized to supply clean electrical power during periods of low solar availability. The integrated DAFC subsystem will be used for this purpose where the stored energy in the form of ammonia will be released through the electrochemical interaction of ammonia molecules as discussed earlier. Moreover, the presented system can also be implemented for other renewable energy resources such as wind power where clean energy is produced intermittently depending on the intensities of wind velocities.

Also, hybrid solar and wind-based plants can also be investigated to be implemented with the presented integrated ammonia synthesis and fuel cell system. The results of the overall energy and exergy efficiencies of the integrated system developed are depicted in Fig. 6.23. The maximum values of efficiencies are observed to occur in the month of July where the energetic efficiency reaches 15.83% considering the overall operation of the system on the monthly average day. In addition, the monthly average day exergetic efficiency is evaluated to be 16.67%. The higher efficiencies for this month may result due to lower amounts of exergy destruction. The exergy analysis of system components is performed considering the ambient temperature for each monthly average day accordingly. Lower exergy destruction amounts lead to lower losses in useful work potential and thus lead to higher overall efficiency. Thus, it is recommended to evaluate the performance of such energy systems according to the respective ambient conditions that vary across the year. Moreover, the variations in efficiencies should also be investigated when other operating parameters including ammonia synthesis pressures, temperatures, WES temperatures, etc., are varied. Such parameters directly affect the required power inputs to the system during the charging phase and hence effect the overall efficiencies. In addition to this, the overall efficiency of the presented system also incorporates the efficiency of the solar PV plant, which entail a maximum efficiency of nearly 15%–17%. Thus, the efficiency of the overall system

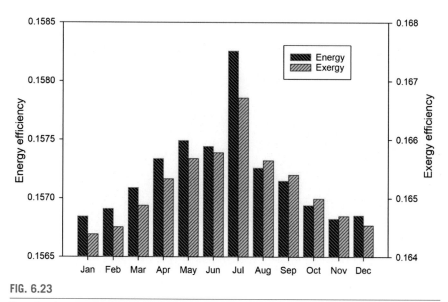

FIG. 6.23

Results of energy and exergy efficiencies on average days of each month.

can also be enhanced through the improvement of solar PV efficiencies. The present system can also be integrated with other types of solar power generation techniques including solar thermal power plants. In addition to this, the power output obtained from the system during the discharging phase can also be increased through the development of high power density DAFC. The power density of the DAFC is an important system parameter that effects the amount of useful electrical power that can be obtained from the system during the discharging phase. Also, higher efficiency DAFC technologies should also be developed that can enhance the efficiencies of the developed system considerably.

The results obtained for the discharge times and AFC energy outputs provided for each monthly average day are depicted in Fig. 6.24. The discharge times are evaluated considering a DAFC peak power density of 6.4 W/m^2 that is associated with a current density of 61.9 A/m^2. The discharge time at the maximum value is observed to be 8.3 h that is found to occur in the month of June. Also, the corresponding energy output provided by the DAFC is evaluated to be 7924.2 kWh. Moreover, the minimum discharge time is evaluated as 2.6 h that is observed to be in the month of December. The energy output during this month was found to be 2492.6 kWh. The discharge time and corresponding energy discharge capacity for each month are directly dependent on the amount of solar radiation intensities during the day, a number of daylight hours, and the corresponding peak power output capacities from the PV plant. In the design of the present system, the excess power threshold is set at 90% of the peak power of each month considered. This parameter can also be

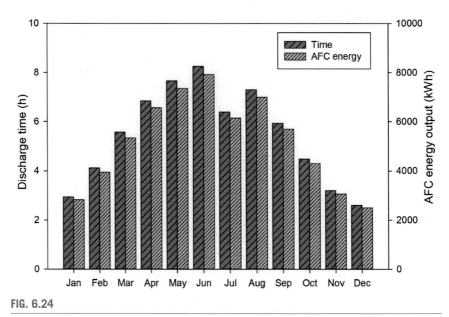

FIG. 6.24

Results of discharge times and AFC energy outputs for monthly average days.

varied and the system performance should be investigated under different thresholds depending on the application as well as the total electrical power output needed from the solar PV plant. At lower cutoff thresholds, a higher amount of ammonia will be synthesized. This will thus lead to higher energy output capacities from the DAFC subsystem. Depending on the given application, the electrical power output from the solar PV plant can be used for the integrated ammonia synthesis and fuel cell system. For instance, one such application could include the usage of the developed system to produce ammonia as an auxiliary fuel for solar PV plants. Conventionally, the electrical energy produced by PV plants is stored in batteries and is later discharged to the grid from the batteries. When the batteries do not entail sufficient energy, auxiliary energy sources such as diesel are used to generate power. However, diesel is a carbon-rich fuel that entails detrimental environmental emissions. Hence, the proposed system can be used to synthesize ammonia during excess power generation, which can be used as an auxiliary source of energy through the DAFC subsystem to generate power when sufficient energy is not available from battery storage.

6.7 Closing remarks

In this chapter, renewable energy-based integrated systems with DAFCs, which have been developed in the recent past are presented. These systems include the opportunity to use the electrochemical interactions of ammonia molecules to produce several useful outputs through system integration. The hybrid DAFC and TES system is firstly presented, which includes the usage of a molten alkaline electrolyte to generate clean power through electrochemical interactions of ammonia molecules as well as stored thermal energy for later usage. The system entails discharging of both electrical as well as thermal energy during the discharging phase. Further, solar- and wind-based integrated energy systems are presented that incorporate the usage of AFCs to produce clean electrical, stored thermal energy as well as electrical energy. Solar thermal power plants entailing excess solar energy are integrated with the hybrid system and their dynamic performances are investigated considering the changes in the solar intensities throughout the year. Also, integrated solar and wind-based energy systems are discussed where the excess energy from both plants is employed for synthesizing ammonia that is used to generate electrical power through DAFCs when there are low wind velocities or solar intensities. The performance of each system is assessed through energy and exergy efficiencies where the total useful energetic and exergetic output is determined as a ratio of the total energetic and exergetic input.

Case studies

7

In this chapter, some novel types of ammonia fuel cell-based technologies are presented, and their performances through energy and exergy efficiencies as well as other performance criteria are investigated to see how they are affected by changing the operating conditions. Comprehensive modeling, analyses, and assessment of each system are discussed to provide insights into the development of new ammonia-based electrochemical systems. The case studies presented provide the detailed thermodynamic analysis equations implemented on each subsystem. Furthermore, the performances of the developed systems are assessed at varying operating conditions and system parameters through different parametric studies. Both the electrochemical, as well as overall performances of the developed systems, are elucidated with recommendations for further improvements.

7.1 Case study 1: Hybrid ammonia fuel cell and battery system with regenerative electrode

The schematic representation depicting the hybrid ammonia fuel cell and battery system with regenerative nickel electrode is given in Fig. 7.1. The developed system entails the operation of both a direct ammonia fuel cell as well as a regenerative electrode battery. The regenerative electrode acts as the common electrode for both the fuel cell as well as the battery subsystem. The system inputs include a portion of ammonia fuel, potassium hydroxide (KOH), and ammonium chloride (NH_4Cl) solutions. During the fuel cell operation, ammonia fuel is input at the catalyst-coated anode where the electrochemical oxidation of ammonia molecules through the interaction with hydroxyl ions occurs. This generates a flow of electrons and thus electricity that is connected to an external circuit where the cathode of the fuel cell comprises the regenerative nickel electrode that absorbs electrons coming from the fuel cell anode. The nickel electrode is regenerated during this process that acts as the anode of the battery subsystem. During battery operation, electrons flow from the nickel anode where oxidation of nickel to positively charged nickel ions occurs. The flow of electrons is accepted at the cathode of the battery, which is immersed in an ammonium chloride medium. Further, the cathodic reaction of the battery produces ammonia molecules that are recycled at the fuel cell anode.

In this way, hybrid fuel cell, as well as battery operation, generates electrical power through a common regenerative nickel electrode. Also, the system includes

Ammonia Fuel Cells. https://doi.org/10.1016/B978-0-12-822825-8.00007-4

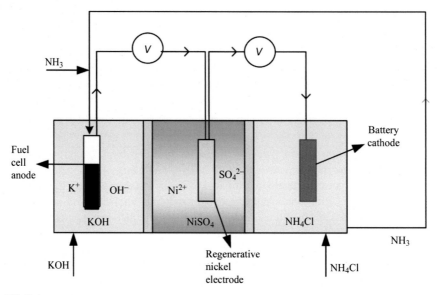

FIG. 7.1

Schematic representation showing the hybrid ammonia fuel cell and battery system with regenerative electrode.

the utilization of ammonium chloride as an electrolyte that produces ammonia during the cathodic reaction of the battery. This ammonia produced is used as the input fuel for the fuel cell operation. The cyclic operation of the hybrid system is repeated with inputs of potassium hydroxide and ammonium chloride to produce the electrical power output. Further, the dual usage of ammonium chloride as an electrolyte as well as a source of ammonia, and the dual usage of the regenerative nickel electrode acting as the fuel cell cathode and battery anode entail the main novelties of the system.

At the fuel cell anode, the input ammonia fuel reacts with negatively charged hydroxyl ions to generate a flow of electrons, however, the positively charged potassium ions remain unreacted. The anolyte of the fuel cell thus comprises a potassium hydroxide solution and the catholyte of the fuel cell is composed of a nickel sulfate solution. Since the solution contains the regenerative electrode, this acts as the catholyte of the fuel cell and the anolyte of the battery. The nickel cations present in this solution are reduced during fuel cell operation and are oxidized during the battery operation, thus entailing regeneration. Moreover, the anode of the fuel cell is composed of a platinum black catalyst to facilitate electrochemical oxidation of ammonia molecules. The cathode of the battery subsystem comprises manganese (IV) oxide material that is immersed in the ammonium chloride catholyte. The anodic half-cell reaction of the fuel cell system is similar to the alkaline electrolyte-based direct ammonia fuel cells discussed earlier, where the electrochemical reaction is denoted as

$$2NH_3 + 6OH^- \rightarrow N_2 + 6H_2O + 6\bar{e} \tag{7.1}$$

Further, the electrons generated at the anode are consumed at the cathode through the following half-cell electrochemical reaction depicting reduction of nickel ions:

$$3Ni^{2+} + 6\bar{e} \rightarrow 3Ni \tag{7.2}$$

These reactions complete the half-cell interactions of the fuel cell subsystem and the overall reaction of the fuel cell can be written as

$$2NH_3 + 6OH^- + 3Ni^{2+} \rightarrow N_2 + 6H_2O + Ni \tag{7.3}$$

The battery operation entails the oxidation of nickel-to-nickel cations that generate the flow of electrons that can be expressed as

$$3Ni \rightarrow 3Ni^{2+} + 6\bar{e} \tag{7.4}$$

Moreover, the electron flow generated at the battery anode is accepted at the cathode to complete the circuit where the cathodic half-cell reaction of the battery can be written as

$$6MnO_2 + 6NH_4Cl + 6\bar{e} \rightarrow 3Mn_2O_3 + 6NH_3 + 3H_2O + 6Cl^- \tag{7.5}$$

Combining both the anodic and cathodic interactions of the battery, the overall reaction can be written as follows for the battery subsystem:

$$3Ni + 6MnO_2 + 6NH_4Cl \rightarrow 3Mn_2O_3 + 6NH_3 + 3Ni^{2+} + 3H_2O + 6Cl^- \tag{7.6}$$

As discussed earlier, the change in Gibbs energy for the fuel cell and battery systems can be evaluated as

$$\Delta G_{FC/BT} = \Delta H_{FC/BT} - T\Delta S_{FC/BT} \tag{7.7}$$

where the subscripts FC and BT denote fuel cell and battery, respectively. Once the change in Gibbs energy is determined, the reversible potential of the electrochemical cell can be found at ambient temperature (T_0) and pressure (P_0) as

$$E^0_{r,FC/BT} = -\frac{\Delta G_{FC/BT(T_0, P_0)}}{nF} \tag{7.8}$$

where the Faraday's constant is denoted by F and the number of electrons is represented by n. As the developed system includes the anolytes and catholytes at varying concentrations, and the reactants and products entail different partial pressures and concentrations, the Nernst equation can be implemented to find the electrical potential as

$$E_{FC} = E_{r,FC} + \frac{RT}{6F} \ln\left(\frac{p_{N_2}}{(p_{NH_3})^2 (C_{OH^-})^6 (C_{Ni^{2+}})^3}\right) \tag{7.9}$$

$$E_{BT} = E_r + \frac{RT}{6F} \ln\left(\frac{(C_{NH_3})^6 (C_{Ni^{2+}})^3 (C_{Cl^-})^6}{(C_{NH_4Cl})^6}\right) \tag{7.10}$$

where the partial pressure of nitrogen is denoted by p_{N_2}, the partial pressure of ammonia is represented by p_{NH_3}, the concentration of hydroxyl ions is written as C_{OH^-}, the concentration of nickel cations is denoted by $C_{Ni^{2+}}$, the chloride ion concentration is written as C_{Cl^-}, the concentration of aqueous ammonia formed is denoted as C_{NH_3}, and the concentration of ammonium chloride is denoted as C_{NH_4Cl}.

These are the fuel cell and battery voltages under open-circuit conditions, however, when an external circuit is connected to the electrochemical cell and current is passed through the circuit, several polarization voltage losses occur in the cell. As discussed earlier, the operating voltage of the electrochemical cell at a given current density can be evaluated as

$$V_{FC/BT} = E_{FC/BT} - V_{act,FC/BT} - V_{ohm,FC/BT} - V_{conc,FC/BT} \tag{7.11}$$

where the terms subtracted from the open-circuit voltage denote the activation, Ohmic, and concentration polarization losses. The evaluation of these losses is similar to the method described in the earlier sections. The activation polarization of the fuel cell and battery subsystems are evaluated according to

$$V_{act,i,FC/BT} = \frac{RT}{\alpha nF} \ln \left(\frac{J_{FC/BT}}{J_{0,i,FC/BT}} \right) \tag{7.12}$$

where the operating current density of the fuel cell or battery system is written as $J_{FC/BT}$, the exchange current density of these subsystems is denoted by $J_{0,i,FC/BT}$ that is considered as 2×10^{-6} mA/cm^2 for the cathodic fuel cell reaction of nickel ion reduction. Also, the anodic exchange current density of the fuel cell subsystem is taken as 1.7×10^{-8} mA/cm^2 and the cathodic exchange current density for the battery subsystem is 0.24 mA/cm^2 [74–76].

The concentration polarization loss in voltage is evaluated for the FC and BT systems according to

$$V_{con,i,FC/BT} = \frac{RT}{nF} \ln \left(\frac{J_{L,i,FC/BT}}{J_{L,i,FC/BT} - J_{FC/BT}} \right) \tag{7.13}$$

where the limiting current density is represented as $J_{L,i,FC/BT}$ and the operating current density is denoted as $J_{FC/BT}$. The limiting current density is calculated as a function of the diffusion coefficient (D), bulk concentration (c_B), and diffusion layer thickness (δ) according to

$$J_{L,i,FC/BT} = nFD_{FC/BT} \frac{C_B}{\delta} \tag{7.14}$$

where the diffusion coefficient values are considered as 6.79×10^{-6} cm^2/s for the diffusion of nickel cations in the catholyte of the fuel cell. Also, the coefficient of diffusion for the diffusion of negatively charged hydroxyl (OH$^-$) ions is taken as 5.27×10^{-5} cm^2/s [77,78]. In addition to this, the thickness of the diffusion layer in the electrolyte is considered to entail a value of 0.03 cm [74]. Further, the voltage loss due to Ohmic polarization in the electrochemical FC or battery subsystems are evaluated as

$$V_{ohm,FC/BT} = J_{FC/BT} R_{FC/BT} \qquad (7.15)$$

where the Ohmic resistance of the electrolytes and other cell components are denoted as $R_{FC/BT}$. Next, the operating current density of the fuel cell system is evaluated in terms of the Faraday's constant as

$$\dot{N}_{NH_3,FC} = \frac{J_{FC}}{nF} \qquad (7.16)$$

The gravimetric energy density is also evaluated for the battery subsystem as follows:

$$En_{d,BT} = \frac{\Delta G_{BT}}{M_{tot}} \qquad (7.17)$$

where the change in Gibb's energy considering the complete battery reaction is represented as ΔG_{BT} and the total molecular mass of the reactants participating in the corresponding reaction is written as M_{tot}. The capacity for energy storage of a given electrode can also be evaluated to assess its capacity. This is evaluated for the regenerative nickel electrode as

$$C_{Ni} = \frac{nF}{3.6 M_{Ni}} \qquad (7.18)$$

where the molecular mass of nickel is represented as M_{Ni} that entails a value of 58.7 g/mol. The power output obtained from either subsystem can be written in terms of the output voltage and current as

$$\dot{W}_{FC/BT} = V_{FC/BT} J_{FC/BT} A \qquad (7.19)$$

where the output cell voltages of either system are written as $V_{FC/BT}$, the output currents are denoted as $J_{FC/BT}$, and the cell area is written as A. The overall performances of the subsytems are evaluated in terms of efficiencies, where the energetic performance of the fuel cell subsystem is evaluated according to

$$\eta_{FC} = \frac{\dot{W}_{FC}}{\dot{N}_{NH_3} \overline{LHV}_{NH_3}} \qquad (7.20)$$

The exergetic performance of the fuel cell system is also assessed that is written as

$$\eta_{ex,FC} = \frac{\dot{W}_{FC}}{\dot{N}_{NH_3} \overline{ex}_{NH_3}} \qquad (7.21)$$

where the power output from the fuel cell is denoted as $\dot{W}_{FC}$, the input molar flow rate of ammonia provided is denoted as $\dot{N}_{NH_3}$, the lower heating value of ammonia is written as $\overline{LHV}_{NH_3}$, and the specific molar exergy of ammonia is represented as $\overline{ex}_{NH_3}$. Next, the battery performance is evaluated in terms of the Nernst voltage ($V_{BT/N}$) and the thermoneutral voltage ($V_{BT/TN}$) as

$$\eta_{BT} = \frac{V_{BT/N}}{V_{BT/TN}} \qquad (7.22)$$

The Nernst battery voltage was evaluated earlier in Eq. (6.10) and the thermoneutral voltage is determined according to the following equation:

$$V_{BT/TN} = -\frac{\Delta h_{BT}}{n_{BT}F} \tag{7.23}$$

where the change in the enthalpy considering the overall battery chemical reaction is denoted as Δh_{BT}, the number of moles of electrons transferred in the electrochemical reactions is represented as n_{BT} and F is the Faraday's constant. The voltage of the battery under the condition of a complete conversion of chemical energy into electrical energy is referred to as the thermoneutral voltage.

The developed system presents a new type of hybrid fuel cell and battery system with the operation of a regenerative nickel electrode. The voltage of the fuel cell system under reversible and open-circuit conditions is evaluated to be 0.54 V considering ambient operating conditions of a temperature of 25°C and pressure of 101 kPa. Moreover, the voltage of the battery cell is determined as 0.73 V under similar reversible, open-circuit and ambient conditions. Next, the limiting current density is determined to entail a value of 1.02 A/cm^2 for the anodic side of the fuel cell. Also, the fuel cell limiting current density at the cathode is evaluated to be 131 mA/cm^2. The limiting current density value for the cathodic side of the battery cell is evaluated as 337.7 mA/cm^2. Moreover, the capacity for energy storage of the nickel regenerative electrode is determined to be 0.92 Ah/g. The gravimetric energy density of the battery cell is also determined to entail a value of 563.1 kJ/kg. Further, the battery cell efficiency in terms of the voltage ratios is determined to be 75%, which denotes the conversion of 75% of the chemical energy stored into electrical energy. The operation of an ideal battery cell where complete conversion of chemical energy into electrical energy occurs, this efficiency value reaches 100%. Moreover, the effects of varying operating parameters on the performance of the developed system are also studied.

The effect of varying operating temperatures on the performance of the battery cell is depicted in Fig. 7.2. The voltage of the cell as a function of the current density at varying temperatures is shown for a temperature range of 25–75°C. Increasing the temperature is observed to be favorable for the battery system where the voltages are found to enhance considerably. For instance, at a given current density value of 100 mA/cm^2, the cell operating voltage rises from nearly 0.51 to 0.67 V when the operating temperature is raised from ambient to 75°C. The open-circuit voltages are also observed to enhance significantly with rising operating temperature. This increase in voltage is directly related to the power output from the battery cell. As can be observed from Eq. (7.19), higher cell voltages at a given current density and cell area result in higher power outputs from the cell. This also leads to higher efficiencies as when the current is kept constant, the input ammonia flow rate is also constant. However, higher power outputs from the same input flow rate result in both higher energy as well as exergy efficiency.

It is thus recommended to implement higher operation temperatures when possible for the developed system. Higher temperature leads to lower polarization

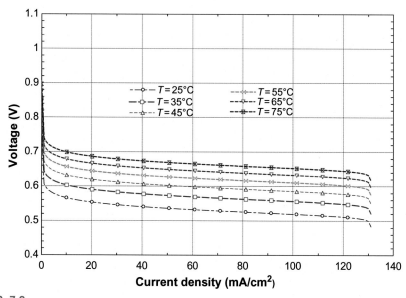

FIG. 7.2

Polarization results for battery cell depicting the operating voltages as a function of current density at different operating temperatures.

losses, this is the primary reason for observing higher battery cell performances in terms of voltage and power output. At higher temperatures, the reactant molecules in the aqueous electrolyte solutions entail higher molecular activity. Hence, at high current densities, the mobility of the molecules entailed in the diffusion layer is more than molecules at lower temperatures. The power density results are shown in Fig. 7.3 where the power output densities increase considerably with temperature. The power densities at the maximum point rise by nearly three times as the temperature rises by 50°C. Therefore, it is recommended to implement higher operating temperatures for the developed system that will aid in achieving higher performances.

The effect of the input pressure of ammonia fuel to the fuel cell system is shown in Fig. 7.4. Rising pressures are observed to enhance the power density values at the maximum point. Before this point, the change in power densities with increasing pressures is significantly lower. Thus, higher input pressures can be used to enhance the power outputs if required, however, it is recommended to investigate the power input needed to pressurize the reactant gases. The input pressure can also be varied from the source tank if ammonia is supplied from a gas tank with regulators. Higher pressures of input ammonia, however, can also lead to lower fuel utilization ratios in direct ammonia fuel cells. As discussed earlier, the catalysts entailing high nitrogen atom adsorption phenomena will be populated with ammonia molecules surrounding

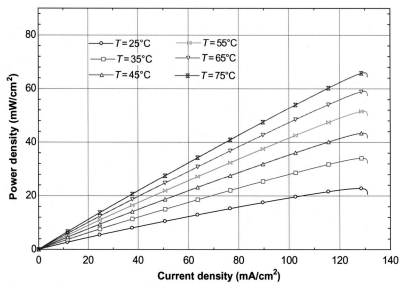

FIG. 7.3

Power density results for battery cell depicting the power densities as a function of current density at different operating temperatures.

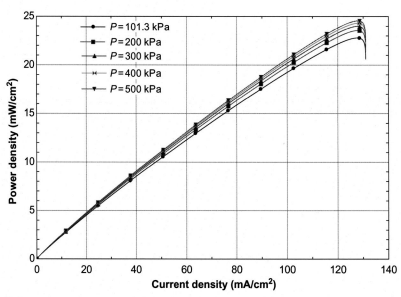

FIG. 7.4

Power density results for fuel cell depicting the power densities as a function of current density at different operating pressures.

the adsorbed species leading to a lower ratio of fuel that reacts electrochemically as compared to the fuel input to the fuel cell. The change in power densities with pressures is generally lower than the change in temperatures.

The effect of varying anolyte concentrations of the fuel cell system on the power output density from the cell is depicted in Fig. 7.5. The aqueous potassium hydroxide solution concentrations are varied from 1 to 5 M and the corresponding power density obtained from the cell is evaluated at varying current densities. The effect of concentration is observed to become more significant at higher current densities. As the current densities rise, the power densities at a given value are observed to be higher for higher anolyte concentrations. At the peak point, the power density is determined to increase by nearly 10 mW/cm^2 as the concentration is raised by five times from 1 to 5 M. In addition to this, the increase in power density with concentration lowers as the concentration reaches higher values. For example, when the concentration is increased from 1 to 2 M, the maximum power density rises from 22.5 to 25 mW/cm^2 that corresponds to an increase of 2.5 mW/cm^2. However, as the concentration increases from 2 to 3 M, the power density rises by 1.5 mW/cm^2. This decreases to nearly 1 mW/cm^2 as the concentrations are increased further from 3 to 4 M and from 4 to 5 M. Hence, the usage of higher concentration anolytes is recommended for thepresent system that will provide higher power densities as well as efficiencies. In addition, it is important to note that in Fig. 7.5 the limiting current density values are considered to be independent of anolyte concentration. However, as the

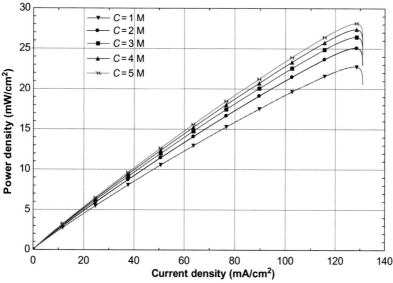

FIG. 7.5

Power density results for fuel cell depicting the power densities as a function of current density at different anolyte concentrations.

concentration increases, the limiting current density also rises as discussed earlier in the analyses. This will lead to a further rise in the power densities. It is recommended to investigate this effect on the system performance, where the diffusion coefficients at these concentrations can be determined and used in the calculation of the limiting current density. The limiting current density will effect the power density calculations leading to higher power densities at higher anolyte concentration. Also, in the anodic fuel cell reactions, the hydroxyl ions are consumed and potassium cations remain unreacted. Hence, it is also recommended to investigate the effect of adding only water to the anolyte as the hydroxyl ions are consumed. Potassium reacts readily with water molecules to form alkaline potassium hydroxide solution. Thus, this could be a system improvement, where the system input will change from being an aqueous potassium hydroxide solution to only water input.

The effect of anolyte concentration on the efficiencies of the fuel cell subsystem is depicted in Fig. 7.6. The efficiencies are observed to follow similar trends as the voltage varies with the current density. At low current densities, activation polarization is considerable that results in significant voltage losses. This is reflected in the efficiency results as can be observed from the figure. Exponential drop is observed at lower current densities in the efficiencies, where both energy as well as exergy efficiencies drop from high values of over 80% to 35%–45% as the current density value reaches $20\,\text{mA/cm}^2$. Higher concentrations entail higher efficiencies both energetically as well as exergetically. Similar to the effect of anolyte concentration on the

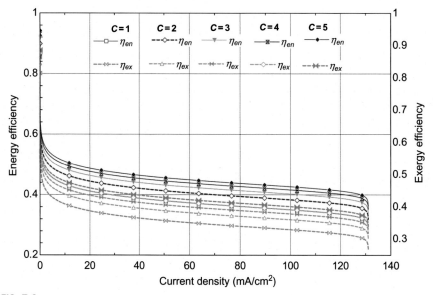

FIG. 7.6

Efficiency results for the fuel cell depicting the energy and exergy efficiencies as a function of current density at different anolyte concentrations.

power density, the effect on the efficiencies becomes more noticeable at higher current densities. Also, the increase in efficiency with concentration decreases with a rise in concentration. These observations are directly attributable to the changes in the power density with concentrations. Rise in power density at a given current density leads to higher power outputs at a given current value and increasing concentrations. Constant current values correspond to constant ammonia fuel inputs and the rise in power output with increasing concentration corresponds to higher useful outputs obtained with the same fuel input. However, other factors including cost, electrode behavior, etc., should also be investigated as a function of the concentration. Higher anolyte concentration will require a higher amount of potassium hydroxide that leads to higher resource costs. Further, the catalyst layers, as well as the electrode materials, may not entail high compatibility with high concentration alkaline anolytes.

The exergy efficiency results for the fuel cell system at varying operating temperatures are depicted in Fig. 7.7. Increasing temperatures are observed to enhance the exergetic efficiencies considerably. The pattern of change with current density is observed following the same trend as the cell voltages. At lower current densities, due to high activation polarization, there are significant voltage losses. These also lead to a drop in the efficiencies as depicted in the figure. This can be attributed to the drop in power output density with a drop in voltage. Lower power outputs for the same fuel input lead to lower values of efficiencies. The exergy efficiency

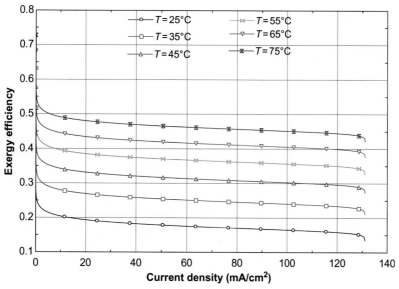

FIG. 7.7

Exergy efficiency results for the fuel cell as a function of current density at different operating temperatures.

depicts the ratio of the useful exergetic output and the provided exergetic input. The output, in this case, comprises the electrical power outputs density that is the product of the output voltage and current density. The exergetic input comprises the product of the exergy content of ammonia fuel per unit mole and the molar input flow rate. Since the current density is a function of the molar input flow rate, the trend observed for the change in efficiencies is similar to the behavior of the output voltage. In addition to this, the exergy efficiency is also related to the amount of exergy destruction occurring within the electrochemical cell. As the exergy efficiency lowers, the exergy destructions are increased. This signifies higher irreversibilities within the cell when the exergy efficiency decreases. The activation polarization region in the lower current density range is the region of a significant drop in exergy efficiency. Thus, efforts need to be directed toward lowering the activation polarization in direct ammonia fuel cells. These can be achieved by increasing the exchange current densities at both the anode as well as cathode. Higher exchange current density signifies the higher rate of electrochemical reactions at equilibrium, which leads to lower activation polarization voltage losses. The initial step where the electrochemical reaction begins is highly dependent on the exchange current densities, this step determines how much loss in voltage and thus efficiencies will occur at low current density values. Better electrochemical catalysts that entail higher compatibility with ammonia molecules and provide higher exchange current densities need to be developed.

The battery subsystem includes nickel anode with nickel sulfate anolyte, manganese oxide cathode with ammonium chloride catholyte. The exergy efficiency of the cell is dependent on the characteristics of each of these components. The concentration of the electrolyte is one such important parameter that effects the performance. This parametric study is depicted in Fig. 7.8, where the output voltage of the battery cell is plotted against the current density at varying anolyte and catholyte concentrations. Fig. 7.8A depicts the voltage change of the battery system with the current density at different ammonium chloride concentrations. Rising concentrations are observed to provide higher voltages at a given current density value. For instance, at a current density of 80 mA/cm^2, the voltage rises from 0.53 V at a concentration of 1 M to 0.57 V at 5 M. This is attributed to the availability of a higher number of reactant ions that are needed to complete that cathodic half-cell reaction. The higher number of reactant molecules lead to the generation of higher voltage potential at a given value of current density. This leads to both higher cell output voltage as well as power densities at a given current density. Higher catholyte concentration can increase the power density by the same magnitude as the voltage is increased. However, the rate of increase in voltage with concentration decreases with rising concentration values. This signifies that lower rise in voltages are observed when the catholyte concentrations are raised further. When the concentration varies from 1 to 2 M, the rise in voltage is higher than when the concentration is increased from 2 to 3 M and further from 3 to 4 M.

The effect of the anolyte concentration on the output voltages is also depicted in Fig. 7.8B where the voltage variation with current density at varying anolyte

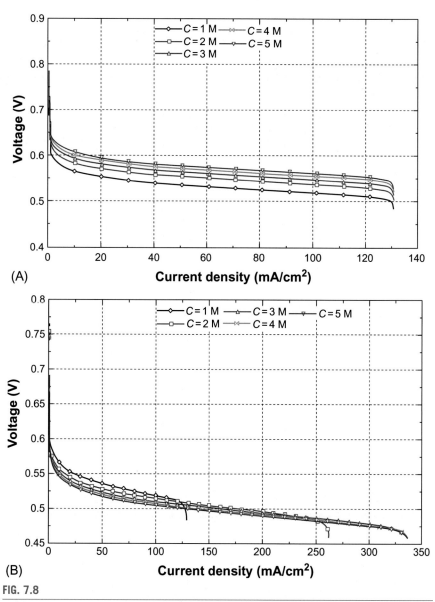

FIG. 7.8

Polarization results for the battery subsystem as a function of current density at different electrolyte concentrations of (A) ammonium chloride and (B) nickel sulfate.

concentrations is shown for a change from 1 to 5 M. In this parametric study, the effect of concentration on limiting current densities is more evident. As the electrolyte concentration increases, the number of moles of reactant ions that participate in the electrochemical reaction increases. Hence, at higher current densities when the diffusion layer becomes significant and molecular diffusion hindrances start occurring leading to limiting currents, higher molar concentration electrolytes entail a higher number of reactants available that can compensate for the missing ions leading to higher limiting currents. As can be observed from the figure, when the concentration increases from 1 to 5 M, the limiting current density rises from nearly 125 to 340 mA/cm^2. The increase in limiting current density is more than 2.7 times, which denotes a significant rise in battery performance. This rise in limiting current densities is directly related to the rise in peak power densities. The output voltages lie in the range of 0.5–0.45 V at the limiting current density values for all molar concentrations. However, as the current density at the limiting value rises by nearly 2.7 times, the power density at the maximum values will also rise by similar magnitudes. The efficiencies, however, depend on the amount of fuel input needed to achieve these current density values. If fuel input rises in proportionality with power output, the change in efficiencies may not be as high as the change in the power densities.

This case study presents a new type of hybrid direct ammonia fuel cell and battery system with a regenerative nickel electrode. The underlying theory of each subsystem along with their corresponding half-cell electrochemical reactions is discussed. The hybrid system entails a regenerative electrode that regenerates during the fuel cell operation and dissociated during the battery operation. The performance of the system is analyzed in terms of the energetic and exergetic assessments along with polarization behaviors. The effects of different operating temperatures, pressures, and electrolyte concentrations are studied for both subsystems. The mean efficiencies of the fuel cell subsystem are evaluated to be 54.2% energetically and 50.9% exergetically. Also, the battery cell entails an efficiency of 75%.

7.2 Case study 2: Integrated solar and geothermal-based system with direct ammonia fuel cells

The integrated solar and geothermal-based system entailing direct ammonia fuel cells is depicted in Fig. 7.9. The solar thermal energy is utilized as the solar energy input and hot geothermal water is employed as the geothermal energy input to the system. Heliostat field reflects the incoming solar radiation onto a central tower receiver where an alkaline molten salt absorbs the incoming energy and storages it in the hybrid molten salt thermal energy storage and direct ammonia fuel cell system (MST). From this subsystem, thermal energy is carried by stream s1 where through heat exchanger HX-1 the thermal energy is transferred to the steam Rankine cycle for power generation. Water at high pressure enters HX-1 and absorbs thermal energy from the incoming molten salt at state s1. The heat transfer process generates high pressure and temperature steam at state 3 that enters the steam turbine to

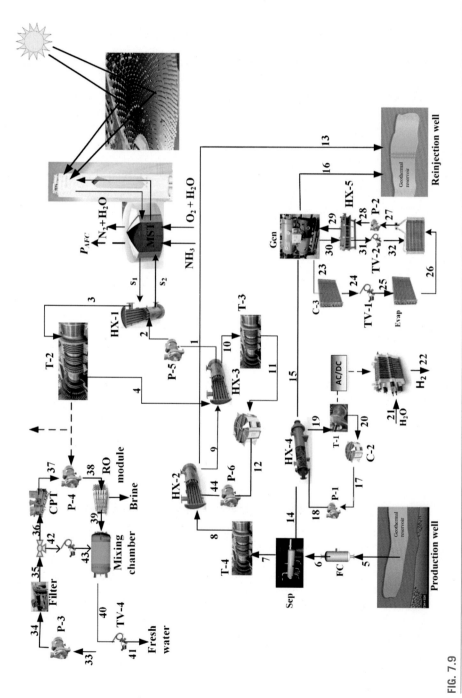

FIG. 7.9

Schematic representation of the integrated solar and geothermal-based multigeneration system with direct ammonia fuel cell.

generate power. A portion of the generated electrical power is transmitted to the reverse osmosis (RO) desalination system for producing fresh water and the remaining is sent to the grid. The RO subsystem operates between states 33 and 43 where salty seawater enters the pump P-3 at state 33 where it is pressurized to a higher pressure before entering the filter. The filter removes the solid and undissolved particles from the saline water that is passed to the chemical pretreatment (CPT) subsystem at state 36. The CPT eradicates the microorganisms that need to be removed to obtain fresh water. The treated water at state 38 is sent to the RO module after being pressurized to the required pressure by pump P-4. The RO module desalinates the water to obtain a stream of salt-free water at state 39. The salt-free water is mixed with a small fraction of saline water coming after the filter stream. The fresh water exits the RO subsystem at state 41 after the appropriate pressure is achieved through throttle valve TV-4. This comprises the second useful output from the multigeneration system after electrical power. Next, the geothermal fluid at high temperature enters the system at state 5 where it is flashed to a lower pressure at state 6 to obtain a higher vapor quality. Dropping the pressure reduces the temperature as well as pressure, however, the vapor quality is increased. This is needed to operate the geothermal steam turbine. The saturated water vapor and liquid mixture enter the separator (SEP) at state 6 where it is separated into vapor and liquid phases at states 7 and 14, respectively. At state 7, the saturated vapor is sent to turbine T-4 where electrical power is generated. The saturated steam exits T-4 at state 8 where it is sent to heat exchanger HX-2 to transfer heat to an organic Rankine cycle (ORC-1) for power generation. The organic fluid is preheated between states 8 and 9. The geothermal fluid is reinjected to the ground after exiting HX-2 at state 13. The ORC-1 operates between states 9 and 12 where pump P-6 pressurizes the organic fluid to a higher pressure. The geothermal fluid enters HX-3 where it absorbs thermal energy from the medium temperature stream 4 that exits T-2. Hot organic fluid exits HX-3 at state 10 and enters turbine T-3 to generate electrical power. After leaving T-3 at state 11, the organic fluid is sent to the condenser where heat is rejected to condense the fluid before entering pump P-6. At state 14, the hot geothermal saturated liquid enters HX-4 where it transfers thermal energy to another ORC-2. In ORC-2, the organic fluid enters the turbine T-1 at a high temperature where electrical power is generated. The power generated from this ORC is utilized to produce hydrogen through PEM water electrolysis. The organic fluid exits T-1 at state 20 and is condensed in C-2 where it enters the pump P-1 at state 17 where it is pressurized to a higher pressure at state 18 before entering HX-4. At state 15, the geothermal fluid still entails high temperatures. This is used to operate an absorption cooling cycle that operates between states 23 and 32. The generator (GEN) of the absorption cooling cycle absorbs thermal energy from the fluid before it is reinjected to the reinjection well at state 16. The absorption cooling absorbs thermal energy from the hot geothermal fluid after entering the GEN at state 29. State 29 comprises a mixture of ammonia and water with a specific mass content. The thermal energy is used by the GEN to separate the ammonia molecules dissolved in water. Ammonia with a high concentration exits GEN at state 23 and enters the condenser C-3 where heat is rejected to cool ammonia. Next, ammonia is passed through the throttle valve TV-1 that drops the pressure as well as the temperature of ammonia before it enters the evaporator (EVAP) to provide cooling. At state 25, cold ammonia enters EVAP where thermal energy is transferred to the fluid to provide cooling to another external fluid. After this process, ammonia exits the evaporator at state 26 and enters the mixing chamber at state 26. In the mixing chamber, high concentration ammonia mixes with low concentration water entering at state 32 to produce a mixed stream that exits the mixing chamber at sate 27. Pump P-2 pressurizes this mixture to a higher pressure at state 28 before it enters HX-5. In HX-5, the high temperature and low concentration stream exiting GEN at state 30 is used to transfer heat to the cold mixed stream coming from the absorber. The preheated mixed stream enters GEN at state 29 where it is separated into water and ammonia streams and the cycle is repeated. In this way, the waste heat entailed in the geothermal fluid is utilized to produce another useful commodity comprising space cooling. Next, the PEM electrolyzer operates with the electrical power generated by T-1. Water enters the subsystem at state 21 and is dissociated into hydrogen and oxygen, where hydrogen exits at state 22.

The developed system is assessed through thermodynamic performance analysis considering steady-state operation. The generic mass balance equation for a component in such systems can be written as

$$\sum_i \dot{m}_{in} = \sum_e \dot{m}_{ex} \tag{7.24}$$

The generic energy balance equation for this steady system is written for a given system component as

$$\dot{Q} - \dot{W} + \sum_i \dot{m}_{in}\left(h_{in} + \frac{V_{in}^2}{2} + gZ_{in}\right) = \sum_{ex} \dot{m}_{ex}\left(h_{ex} + \frac{V_{ex}^2}{2} + gZ_{ex}\right) \tag{7.25}$$

Next, the entropy balance is performed for each component of the system, where the general equation is expressed as

$$\sum_i \dot{m}_{in}s_{in} + \dot{S}_{gen} + \sum_k \frac{\dot{Q}_k}{T_k} = \sum_{ex} \dot{m}_{ex}s_{ex} \tag{7.26}$$

Also, the exergy balance is performed according to

$$\dot{Ex}^Q + \sum_i \dot{m}_{in}ex_{in} = \sum_{ex} \dot{m}_{ex}ex_{ex} + \dot{Ex}_{wk} + \dot{Ex}_{dt} \tag{7.27}$$

Several assumptions are made in the present analysis to facilitate the simulation of the system. First, the operation of mechanical components including turbines, pumps, valves, etc., are considered to entail adiabatic form of operation where no heat losses to the surroundings occur during operation. Also, the potential energy changes due to elevation differences are assumed to be negligible. Further, the heat exchangers are assumed to operate with no losses in fluid pressures during operation. The operating isentropic efficiency of mechanical system components including turbines and pumps is considered to be 85%. The geothermal well properties considered for the analysis are a well temperature of 240°C where the fluid extracted is in a saturated liquid state. The saturated liquid is flashed to a lower pressure that increases the vapor quality. The inlet pressure at the flash chamber is 3347 kPa and the flash pressure comprises 1000 kPa. The vapor quality obtained through this flashing process is 0.137. The vapor separated in the separator is sent to the turbine at the flash pressure and the corresponding saturation temperature. The turbine operates with an isentropic efficiency of 85%. The saturated liquid portion of the fluid at the flash pressure is sent to HX-4 where heat is transferred to the ORC operating with n-octane as the organic fluid. At states 17 and 12, the organic fluid is considered to be in the saturated liquid state. The thermodynamic balance equations for the flash chamber and separator are written, respectively, as follows:

Flash chamber:

$$\dot{m}_5 = \dot{m}_6 \tag{7.28}$$

$$\dot{m}_5 h_5 = \dot{m}_6 h_6 \tag{7.29}$$

$$\dot{m}_5 s_5 + \dot{S}_{gen,FC} = \dot{m}_6 s_6 \tag{7.30}$$

$$\dot{m}_5 ex_5 = \dot{m}_6 ex_6 + \dot{E}x_{dest,FC} \tag{7.31}$$

Separator:

$$\dot{m}_6 = \dot{m}_7 + \dot{m}_{14} \tag{7.32}$$

$$\dot{m}_6 h_6 = \dot{m}_7 h_7 + \dot{m}_{14} h_{14} \tag{7.33}$$

$$\dot{m}_6 s_6 + \dot{S}_{gen.sep} = \dot{m}_7 s_7 + \dot{m}_{14} s_{14} \tag{7.34}$$

$$\dot{m}_6 ex_6 = \dot{m}_7 ex_7 + \dot{m}_{14} ex_{14} + \dot{E}x_{dest,sep} \tag{7.35}$$

Next, the turbine T-4 generates useful electrical power and the balance equations for this component are expressed as

$$\dot{m}_7 = \dot{m}_8 \tag{7.36}$$

$$\dot{m}_7 h_7 = \dot{W}_{T4} + \dot{m}_8 h_8 \tag{7.37}$$

$$\dot{m}_7 s_7 + \dot{S}_{gen,T4} = \dot{m}_8 s_8 \tag{7.38}$$

$$\dot{m}_7 ex_7 = \dot{m}_8 ex_8 + \dot{W}_{T4} + \dot{E}x_{dest,T4} \tag{7.39}$$

Further, the hot saturated steam enters HX-2 to transfer thermal energy to the ORC. The balance equations for HX-2 are written as

$$\dot{m}_8 = \dot{m}_{13} \text{ and } \dot{m}_{44} = \dot{m}_9 \tag{7.40}$$

$$\dot{m}_8 h_8 + \dot{m}_{44} h_4 = \dot{m}_9 h_9 + \dot{m}_{13} h_{13} \tag{7.41}$$

$$\dot{m}_8 s_8 + \dot{m}_{44} s_{44} + \dot{S}_{gen,HX2} = \dot{m}_9 s_9 + \dot{m}_{13} s_{13} \tag{7.42}$$

$$\dot{m}_8 ex_8 + \dot{m}_{44} ex_{44} = \dot{m}_9 ex_9 + \dot{m}_{13} ex_{13} + \dot{E}x_{dest,HX2} \tag{7.43}$$

The ORC-1 pump P-6 is also analyzed thermodynamically where the balance equations can be written as

$$\dot{m}_{12} = \dot{m}_{44} \tag{7.44}$$

$$\dot{m}_{12}h_{12} + \dot{W}_{P6} = \dot{m}_{44}h_{44} \tag{7.45}$$

$$\dot{m}_{12}s_{12} + \dot{S}_{gen,P6} = \dot{m}_{44}s_{44} \tag{7.46}$$

$$\dot{m}_{12}ex_{12} + \dot{W}_{P6} = \dot{m}_{44}ex_{44} + \dot{Ex}_{dest,P6} \tag{7.47}$$

The organic fluid absorbs thermal energy from both HX-2 as well as HX-3, where HX-3 is analyzed as

$$\dot{m}_9 = \dot{m}_{10} \text{ and } \dot{m}_4 = \dot{m}_1 \tag{7.48}$$

$$\dot{m}_9h_9 + \dot{m}_4h_4 = \dot{m}_{10}h_{10} + \dot{m}_1h_1 \tag{7.49}$$

$$\dot{m}_9s_9 + \dot{m}_4s_4 + \dot{S}_{gen,HX3} = \dot{m}_{10}s_{10} + \dot{m}_1s_1 \tag{7.50}$$

$$\dot{m}_9ex_9 + \dot{m}_4ex_4 = \dot{m}_{10}ex_{10} + \dot{m}_1ex_1 + \dot{Ex}_{dest,HX3} \tag{7.51}$$

Further, the ORC turbine T-3 generates power, which is analyzed according to

$$\dot{m}_{10} = \dot{m}_{11} \tag{7.52}$$

$$\dot{m}_{10}h_{10} = \dot{W}_{T3} + \dot{m}_{11}h_{11} \tag{7.53}$$

$$\dot{m}_{10}s_{10} + \dot{S}_{gen,T3} = \dot{m}_{11}s_{11} \tag{7.54}$$

$$\dot{m}_{10}ex_{10} = \dot{m}_{11}ex_{11} + \dot{W}_{T3} + \dot{Ex}_{dest,T3} \tag{7.55}$$

Next, the condenser of the ORC rejects heat and the organic fluid n-octane is condensed before entering P-6. The balance equations for the ORC condenser are written as

$$\dot{m}_{11} = \dot{m}_{12} \tag{7.56}$$

$$\dot{m}_{11}h_{11} = \dot{m}_{12}h_{12} + \dot{Q}_{Con} \tag{7.57}$$

$$\dot{m}_{11}s_{11} + \dot{S}_{gen,Con} = \dot{m}_{12}s_{12} + \frac{\dot{Q}_{Con}}{T} \tag{7.58}$$

$$\dot{m}_{11}ex_{11} = \dot{m}_{12}ex_{12} + \dot{Q}_{Con}\left(1 - \frac{T_0}{T_{con}}\right) + \dot{Ex}_{dest,Con} \tag{7.59}$$

In this way, the ORC-1 is analyzed with the thermodynamic assessment of each system component. Moreover, the ORC-2 that operates between states 17 and 20 also operates in a similar way where the ORC pump P-1 is analyzed as

$$\dot{m}_{17} = \dot{m}_{18} \tag{7.60}$$

$$\dot{m}_{17}h_{17} + \dot{W}_{P1} = \dot{m}_{18}h_{18} \tag{7.61}$$

$$\dot{m}_{17}s_{17} + \dot{S}_{gen,P1} = \dot{m}_{18}s_{18} \tag{7.62}$$

$$\dot{m}_{17}ex_{17} + \dot{W}_{P1} = \dot{m}_{18}ex_{18} + \dot{Ex}_{dest,P1} \tag{7.63}$$

In addition to this, the ORC-2 obtains its input thermal energy from HX-4 where the hot saturated liquid provides the required energy. The thermodynamic analysis of HX-4 is performed according to

$$\dot{m}_{14} = \dot{m}_{15} \text{ and } \dot{m}_{18} = \dot{m}_{19} \tag{7.64}$$

$$\dot{m}_{14}h_{14} + \dot{m}_{18}h_{18} = \dot{m}_{19}h_{19} + \dot{m}_{15}h_{15} \tag{7.65}$$

$$\dot{m}_{14}s_{14} + \dot{m}_{18}s_{18} + \dot{S}_{gen,HX4} = \dot{m}_{19}s_{19} + \dot{m}_{15}s_{15} \tag{7.66}$$

$$\dot{m}_{14}ex_{14} + \dot{m}_{18}ex_{18} = \dot{m}_{19}ex_{19} + \dot{m}_{15}ex_{15} + \dot{E}x_{dest,HX4} \tag{7.67}$$

Moreover, the ORC turbine T-1 is assessed thermodynamically according to

$$\dot{m}_{19} = \dot{m}_{20} \tag{7.68}$$

$$\dot{m}_{19}h_{19} = \dot{W}_{T1} + \dot{m}_{20}h_{20} \tag{7.69}$$

$$\dot{m}_{19}s_{19} + \dot{S}_{gen,T1} = \dot{m}_{20}s_{20} \tag{7.70}$$

$$\dot{m}_{19}ex_{19} = \dot{m}_{20}ex_{20} + \dot{W}_{T1} + \dot{E}x_{dest,T1} \tag{7.71}$$

After exiting T-1, the organic fluid is passed through condenser C-2. The balance equations for this component are

$$\dot{m}_{20} = \dot{m}_{17} \tag{7.72}$$

$$\dot{m}_{20}h_{20} = \dot{m}_{17}h_{17} + \dot{Q}_{C-2} \tag{7.73}$$

$$\dot{m}_{20}s_{20} + \dot{S}_{gen,C-2} = \dot{m}_{17}s_{17} + \frac{\dot{Q}_{C-2}}{T} \tag{7.74}$$

$$\dot{m}_{20}ex_{20} = \dot{m}_{17}ex_{17} + \dot{Q}_{C-2}\left(1 - \frac{T_0}{T_{c-2}}\right) + \dot{E}x_{dest,C-2} \tag{7.75}$$

The absorption cooling cycle obtains the required thermal energy input from the hot saturated liquid at state 15. The balance equation for the generator of this cycle can be written as

$$\dot{m}_{29} = \dot{m}_{23} + \dot{m}_{30} \tag{7.76}$$

$$\dot{m}_{29}h_{29} + \dot{Q}_{gen} = \dot{m}_{23}h_{23} + \dot{m}_{30}h_{30} \tag{7.77}$$

$$\dot{m}_{29}s_{29} + \frac{\dot{Q}_{gen}}{T_{gen}} + \dot{S}_{gen} = \dot{m}_{23}s_{23} + \dot{m}_{30}s_{30} \tag{7.78}$$

$$\dot{m}_{29}ex_{29} + \dot{Q}_{gen}\left(1 - \frac{T_0}{T}\right) = \dot{m}_{23}ex_{23} + \dot{m}_{30}ex_{30} \tag{7.79}$$

The thermodynamic analyses of the absorption cooling subsystem condenser C-3 can be expressed as

$$\dot{m}_{23} = \dot{m}_{24} \tag{7.80}$$

$$\dot{m}_{23}h_{23} = \dot{Q}_{l,C-3} + \dot{m}_{24}h_{24} \tag{7.81}$$

$$\dot{m}_{23}s_{23} + \dot{S}_{gen,C-3} = \dot{m}_{24}s_{24} + \frac{\dot{Q}_{l,C-3}}{T_0} \tag{7.82}$$

$$\dot{m}_{23}ex_{23} = \dot{m}_{24}ex_{20} + \dot{Q}_{l,con}\left(1 - \frac{T_0}{T_{con}}\right) + \dot{Ex}_{d,con} \tag{7.83}$$

Ammonia enters throttle valve TV-1 after leaving C-3, where the pressure and thus temperature are dropped. The analysis of this component is expressed as

$$\dot{m}_{24} = \dot{m}_{25} \tag{7.84}$$

$$\dot{m}_{24}h_{24} = \dot{m}_{25}h_{25} \tag{7.85}$$

$$\dot{m}_{24}s_{24} + \dot{S}_{gen,TV1} = \dot{m}_{25}s_{25} \tag{7.86}$$

$$\dot{m}_{24}ex_{24} = \dot{m}_{25}ex_{25} + \dot{Ex}_{d,TV1} \tag{7.87}$$

The evaporator is the component where the actual cooling process takes place. The balance equations for this component are

$$\dot{m}_{25} = \dot{m}_{26} \tag{7.88}$$

$$\dot{m}_{25}h_{25} + \dot{Q}_{Evap} = \dot{m}_{26}h_{26} \tag{7.89}$$

$$\dot{m}_{25}s_{25} + \frac{\dot{Q}_{Evap}}{T_0} + \dot{S}_{gen,Evap} = \dot{m}_{26}s_{26} \tag{7.90}$$

$$\dot{m}_{25}ex_{25} + \dot{Q}_{Evap}\left(1 - \frac{T_0}{T_{Evap}}\right) = \dot{m}_{26}ex_{26} + \dot{Ex}_{d,Evap} \tag{7.91}$$

Next, the ammonia is passed to the absorber system component. The thermodynamic analyses of this system component are denoted as

$$\dot{m}_{26} + \dot{m}_{32} = \dot{m}_{27} \tag{7.92}$$

$$\dot{m}_{26}h_{26} + \dot{m}_{32}h_{32} = \dot{m}_{27}h_{27} + \dot{Q}_{abs} \tag{7.93}$$

$$\dot{m}_{26}ex_{26} + \dot{m}_{32}ex_{32} = \dot{m}_{27}ex_{27} + \dot{Q}_{abs}\left(1 - \frac{T_0}{T_{abs}}\right) \tag{7.94}$$

The system performance is assessed in terms of the overall energetic and exergetic efficiencies. The performance of the ORC-1 is evaluated as

$$\eta_{ORC-1} = \frac{\dot{W}_{T-3}}{\dot{m}_5(h_5 - h_4) + \dot{m}_6(h_6 - h_9)} \tag{7.95}$$

$$\eta_{ex,ORC-1} = \frac{\dot{W}_{T-3}}{\dot{m}_5(ex_5 - ex_4) + \dot{m}_6(ex_6 - ex_9)} \tag{7.96}$$

Also, the energetic and exergetic efficiencies of the ORC-2 can be written as

$$\eta_{ORC-2} = \frac{\dot{W}_{T-4}}{\dot{m}_{17}(h_{16} - h_{17})} \tag{7.97}$$

$$\eta_{ex,ORC-2} = \frac{\dot{W}_{T-4}}{\dot{m}_{17}(ex_{16} - ex_{17})} \tag{7.98}$$

The energetic and exergetic efficiencies of the geothermal power generation is evaluated according to

$$\eta_{geo} = \frac{\dot{W}_{T-1}}{\dot{m}_1 h_1 - \dot{m}_5 h_5 - \dot{m}_{18} h_{18}} \tag{7.99}$$

$$\eta_{ex,geo} = \frac{\dot{W}_{T-1}}{\dot{m}_1 ex_1 - \dot{m}_5 ex_5 - \dot{m}_{18} ex_{18}} \tag{7.100}$$

The solar tower-based power generation efficiencies integrated with the direct ammonia fuel cell are evaluated as

$$\eta_{solar} = \frac{\dot{W}_{T-2} + \dot{P}_{AFC}}{\dot{I}_{sol} N_h A_h + \dot{m}_{NH_3} LHV_{NH_3}} \tag{7.101}$$

$$\eta_{ex,solar} = \frac{\dot{W}_{T-2} + \dot{P}_{AFC}}{\dot{I}_{sol} N_h A_h \left(1 - \dfrac{T_0}{T_{sun}}\right) + \dot{m}_{NH_3} ex_{NH_3}} \tag{7.102}$$

The RO desalination subsystem performance is determined as

$$\eta_{RO} = \frac{\dot{m}_{43} h_{43}}{\dot{W}_{P-3} + \dot{W}_{P-4} + \dot{m}_{35} h_{35}} \tag{7.103}$$

$$\eta_{ex,RO} = \frac{\dot{m}_{43} ex_{43}}{\dot{W}_{P3} + \dot{W}_{P4} + \dot{m}_{35} ex_{35}} \tag{7.104}$$

Moreover, the performance of the ammonia fuel cell subsystem is evaluated according to

$$\eta_{AFC} = \frac{\dot{P}_{AFC}}{\dot{m}_{NH_3} LHV_{NH_3}} \tag{7.105}$$

$$\eta_{ex,AFC} = \frac{\dot{P}_{AFC}}{\dot{m}_{NH_3} ex_{NH_3}} \tag{7.106}$$

Further, the hydrogen production energetic and exergetic efficiencies are evaluated as

$$\eta_{H_2} = \frac{\dot{m}_{H_2} LHV_{H_2}}{\dot{P}_{in,PEM}} \tag{7.107}$$

$$\eta_{ex,H_2} = \frac{\dot{m}_{H_2} ex_{H_2}}{\dot{P}_{in,PEM}} \tag{7.108}$$

The energetic, as well as exergetic performance of the absorption cooling subsystem, is determined in terms of the coefficient of performance (COP):

$$COP_{en} = \frac{\dot{Q}_{EVAP}}{\dot{Q}_{gen}} \tag{7.109}$$

$$COP_{ex} = \frac{\dot{Q}_{EVAP}\left(\frac{T_0}{T_{EVAP}} - 1\right)}{\dot{Q}_{gen}\left(1 - \frac{T_0}{T_{gen}}\right)} \tag{7.110}$$

The overall performance of the developed system in terms of energy efficiency is evaluated as

$$\eta_{en} =$$

$$\frac{\dot{W}_{T-1} + \dot{W}_{T-2} + \dot{W}_{T-3} - \dot{W}_{P-1} - \dot{W}_{P-2} - \dot{W}_{P-3} - \dot{W}_{P-4} - \dot{W}_{P-5} - \dot{W}_{P-6} + \dot{Q}_{Evap} + \dot{m}_{43}h_{43} + \dot{P}_{AFC} + \dot{m}_{H_2}LHV_{H_2}}{\dot{I}_{sol}N_hA_h + (\dot{m}_1h_1 - \dot{m}_5h_5 - \dot{m}_{18}h_{18}) + \dot{m}_{35}h_{35} + \dot{m}_{NH_3}LHV_{NH_3}}$$

$$\tag{7.111}$$

Also, the overall exergy efficiency is calculated as

$$\eta_{ex} =$$

$$\frac{\dot{W}_{T-1} + \dot{W}_{T-2} + \dot{W}_{T-3} - \dot{W}_{P-1} - \dot{W}_{P-2} - \dot{W}_{P-3} - \dot{W}_{P-4} - \dot{W}_{P-5} - \dot{W}_{P-6} + \dot{Q}_{EVAP}\left(\frac{T_0}{T_{EVAP}} - 1\right) + \dot{m}_{43}ex_{43} + \dot{P}_{AFC} + \dot{m}_{H_2}ex_{H_2}}{\dot{I}_{sol}N_hA_h\left(1 - \frac{T_0}{T_{sun}}\right) + (\dot{m}_1ex_1 - \dot{m}_5ex_5 - \dot{m}_{18}ex_{18}) + \dot{m}_{35}ex_{35} + \dot{m}_{NH_3}ex_{NH_3}}$$

$$\tag{7.112}$$

The thermodynamic analyses of the developed system are performed through the engineering equation solver. The system parameters used in the analyses and thermodynamic properties of system state points are provided in Tables 7.1 and 7.2, respectively. Moreover, the rates of exergy destruction and associated subsystems efficiencies are summarized in Tables 7.3 and 7.4, respectively. The overall energy efficiency of the developed system is found to be 42.3% and the overall exergy efficiency is determined as 21.3%. Moreover, the energy efficiency of ORC 1 and ORC 2 are evaluated as 8.1% and 11.7%, respectively. However, both were found to entail high exergetic efficiencies of 53.8% and 50.4%, respectively. In addition, the absorption cooling cycle is determined to have an energetic COP of 0.54. Also, the exergetic COP is evaluated to be 0.31. The geothermal and solar-based power generation subsystems are found to be 25% and 18.4% efficient energetically. Also, they entail exergy efficiencies of 41.7% and 19.2%, respectively. The integrated multigeneration system thus aids in achieving higher energy efficiencies than single power generation from geothermal or solar power plants. The exergy efficiency of the overall system is higher than the solar power generation with AFC and lower than the geothermal-based power generation. The solar integrated AFC is determined to have an energetic efficiency of 23.2% and an exergetic efficiency of 21.7%. The RO desalination subsystem entails efficiencies of 62.9% and 29.7% energetically and exergetically, respectively. Further, the absorption cooling cycle generator and the geothermal flash chamber are found to have comparatively higher exergy destruction rates of 2370.2 and 643.3 kW. These can be attributed to high temperature differences and changes. Also, turbine T-2 is evaluated to have comparatively higher exergy destruction rate of 341 kW. Also, the turbine T-1 entails a rate of exergy destruction of 261.3 kW. These can be reduced through usage of turbines that are associated with high isentropic efficiencies. Several parametric studies are also performed to analyze how the performances vary as the operating conditions are changed.

Table 7.1 System parameters and operating conditions used in the analyses.

Parameter	Value
Solar radiation intensity	$850\,W/m^2$
Number of heliostats	200
Mirror (heliostat) dimensions	$11\,m \times 11\,m$
Efficiency of heliostat field	85%
Type of water splitting	Proton exchange membrane (PEM)
Membrane thickness	$100\,\mu m$
PEM temperature	25°C
PEM preexponential factors	Anodic: $1.7 \times 10^5\,A/m^2$
	Cathodic: $4.6 \times 10^3\,A/m^2$
Faraday's constant	$96{,}485\,C/mol$
PEM operating pressure	$101.1\,kPa$
Geothermal fluid temperature	240°C
Geothermal fluid pressure	$3347\,kPa$
Geothermal fluid mass flow rate	$50\,kg/s$
Geothermal flashing pressure	$1000\,kPa$
Molten salt electrolyte	49 mol% KOH and 51 mol% NaOH
Ammonia fuel cell exchange current density	$370\,mA/m^2$
Ammonia fuel cell limiting current density	$20{,}000\,A/m^2$
Molten electrolyte temperature	400°C
Salinity of RO inlet	$35{,}000\,ppm$
Salinity of desalinated water	$450\,ppm$
Membrane salt rejection	90%
Membrane recovery ratio	0.6
Absorption cooling cycle working fluid	NH_3-H_2O mixture
Absorption cooling cycle high pressure	$1555.8\,kPa$
Absorption cooling cycle low pressure	$244.9\,kPa$
Absorption cooling cycle concentrated stream ammonia mass fraction	0.99634
Turbine and pump isentropic efficiencies	85%

Increasing geothermal fluid temperatures are observed to result in higher overall efficiencies. This is depicted in Fig. 7.10. For instance, with an increase in the geothermal fluid temperature from 200 to 250°C, the energetic efficiency is observed to rise from 40.9% to 42.7% and the exergetic performance is observed to rise from an exergetic efficiency of 17.8% to 22.1%. This is attributed to an increase in the overall useful outputs of the system. In addition to this, the efficiencies of geothermal power generation are also observed to rise with the fluid temperature. As the geothermal fluid temperature rises from 200 to 250°C, the geothermal energy efficiency rises from 9.8% to 28.2%. Also, the geothermal exergy efficiency increases from 24.3% to 43.3% for the same temperature rise. However, the exergy destruction rate in flash chamber is also observed to rise with rising temperatures. For the same

Table 7.2 Thermophysical properties of system state points.

State no.	Working fluid	Pressure (kPa)	Tempera-ture (°C)	Mass flow rate (kg/s)	Specific exergy (kJ/kg K)	Specific Enthalpy (kJ/kg)
1	Water	3347	240	50	236.8	1037
2	Water	1000	179.9	50	224	1037
3	Water	1000	179.9	6.8	819.3	2777
4	Water	150	111.3	6.8	501.1	2497
5	Water	150	111.3	6.8	472.8	2372
6	Water	75	91.7	5	27.4	384.4
7	Water	1000	91.8	5	28.4	385.3
8	Water	1000	681.6	5	1434	3882
9	Water	75	332.2	5	623.3	3140
10	n-Octane	15.9	70	20	7.15	104
11	n-Octane	1800	70	20	9.69	15.5
12	n-Octane	1800	87.6	20	16.3	148.3
13	n-Octane	1800	88.7	20	15.2	141.9
14	n-Octane	15.9	70	20	11.8	139
16	Water	1000	179.9	43.2	129.9	762.5
17	Water	1000	168.3	43.2	112.7	711.7
18	Water	1000	154	43.2	93.5	649.6
19	n-Octane	13.2	65.5	8.5	5.79	93.2
20	n-Octane	1900	65.6	8.5	8.49	94.8
21	n-Octane	1900	165	8.5	68.2	353.1
22	n-Octane	13.2	65.5	8.5	33.4	322.9
23	Ammonia/water	1556	107.4	3.12	401.1	1545
24	Ammonia/water	1556	41.1	3.12	317.5	189.8
25	Ammonia/water	245	−13.98	3.12	286.9	189.8
26	Ammonia/water	245	−9.87	3.12	129.8	1257.9
27	Ammonia/water	245	38.9	22.1	−7.29	−44.4
28	Ammonia/water	1556	39.3	22.1	−5.71	−41.3
29	Ammonia/water	1556	109.5	22.1	44.9	301.2
30	Ammonia/water	1556	129.3	19.08	67.8	394.8
31	Ammonia/water	1556	39.3	19.08	2.95	−0.81
32	Ammonia/water	245	40.2	19.08	1.56	−0.81
35	Seawater	101	44.5	145.4	2.58	178.7
36	Seawater	652	44.6	145.4	2.62	179.2
37	Seawater	629	44.6	145.4	2.62	179.2
38	Seawater	629	44.6	145.2	2.62	179.2
39	Saline water	610	44.6	145.2	2.62	179.2
40	Saline water	6002	46.2	145.2	3.02	185.5

Continued

Table 7.2 Thermophysical properties of system state points—cont'd

State no.	Working fluid	Pressure (kPa)	Temperature (°C)	Mass flow rate (kg/s)	Specific exergy (kJ/kg K)	Specific Enthalpy (kJ/kg)
41	Desalinated water	111	46.2	87	4.62	196.6
42	Fresh water	111	45.1	87.25	4.33	191.9
43	Fresh water	101	46.2	87.25	4.63	196.5
44	Saline water	629	44.6	0.25	2.62	179.2
45	Saline water	111	44.6	0.25	2.63	179.2

Table 7.3 Results for major exergy destruction rates in the developed system.

System component	Exergy destruction rate (kW)
Flash chamber	643.3
Heat exchanger HX-1	4814
Heat exchanger HX-4	232.2
Heat exchanger HX-5	120.2
Turbine T-1	261.3
Turbine T-2	341
Turbine T-3	9.02
Turbine T-4	39.75
Condenser C-2	1.2
Condenser C-3	10.53
Evaporator	932.4
Generator	2370.3
RO module	9.22

Table 7.4 Results of energy and exergy efficiencies of subsystems and overall system.

System	Energy efficiency	Exergy efficiency
Organic Rankine cycle 1	8.1%	53.8%
Organic Rankine cycle 2	11.7%	50.4%
Geothermal power generation	25%	41.7%
PEM electrolysis	46.8%	45.3%
Solar power generation with AFC	18.4%	19.2%
Ammonia fuel cell	23.2%	21.7%
RO desalination	62.9%	29.7%
Absorption cooling cycle	0.541 (COP)	0.31 (COP)
Overall system	42.3%	21.3%

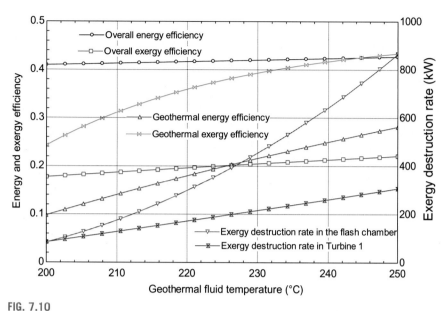

FIG. 7.10

Effect of geothermal fluid temperature on the overall efficiencies and exergy destruction rates.

temperature rise described above, the exergy destruction rate rises from 84.3 to 854.9 kW. This can be attributed to the rise in temperature at the fixed flashing pressure. Thus, it is suggested to use optimal flashing temperatures and the corresponding flash pressures.

The effects of varying direct normal irradiance on the overall efficiencies, turbine T-2 power output, and exergy destruction rate is depicted in Fig. 7.11. Increasing solar irradiance is observed to result in rising overall exergy efficiency. However, energy efficiency is observed to decrease with rising solar irradiance. At a solar irradiance of 0.29 kW/m^2, the overall energy efficiency is found to be 49% and the overall exergetic efficiency is observed to be 20.3%. However, at a high intensity of 900 W/m^2, the overall energy efficiency reduces to 42.1% and the overall exergy efficiency increases to 21.8%. Also, as can be observed from Fig. 3, after an irradiance value of 0.6 kW/m^2 the trends are observed to vary. The overall exergy efficiency reduces until the solar irradiance reaches this value and then starts increasing. Further, the power output of the turbine T-2 rises more steadily with increase in irradiance after 0.6 kW/m^2. Thus, it is recommended to analyze the overall performances of solar-based integrated systems at varying solar irradiances to determine the performance of the plant at varying intensities.

The voltage and power density vs current density curves for the ammonia fuel cell are depicted in Fig. 7.12 for different operating temperatures. The peak power density is observed to be nearly 2900 W/m^2 at an operating temperature of 500°C. This is observed to decrease with decreasing temperatures. For instance, at electrolyte

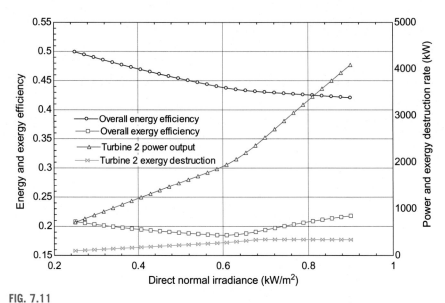

FIG. 7.11

Effect of direct normal irradiance on the overall efficiencies and exergy destruction rates.

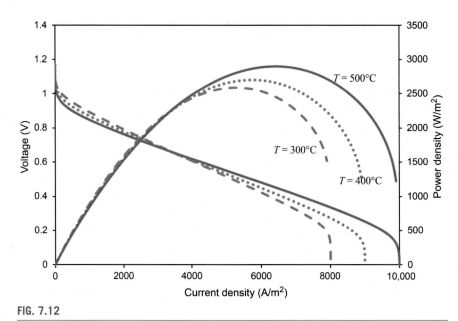

FIG. 7.12

Voltage and power vs current density curves for molten alkaline electrolyte ammonia fuel cell.

temperatures of 400 and 300°C, power densities of nearly 2691 and 2585 W/m^2 are observed, respectively. The increase in fuel cell performance with temperature can be attributed to the enhancement in the electrolyte conductivity at high temperatures. Higher conductivities lead to lower Ohmic as well as concentration polarization losses due to higher molecular activity. Hence, usage of optimal operating conditions is recommended, which provide both higher power outputs as well as efficiencies.

Fig. 7.13 depicts the effects of turbine isentropic efficiencies on system performance. The overall energetic system performance in terms of energy efficiency rises from 40.2% to 42.3% when the isentropic turbine efficiency is increased from 65% to 85%. Also, the exergetic performance enhances from an exergy efficiency of 16.2% to 21.3%. This can be attributed to the rise in useful power as well as drop in irreversibilities and exergy destruction rates. The efficiencies of subsystems are also observed to increase with rising isentropic efficiencies. For instance, the energy efficiencies of ORC-1, ORC-2, solar, and geothermal power generation rise from 6.2%, 8.9%, 14.5%, and 20.3% to 8.1%, 11.7%, 18.4%, and 25%, respectively, for an isentropic efficiency rise from 65% to 85%. Similarly, the exergetic performances are also observed to increase considerably for the associated subsystems. For instance, the ORC-1 entails an increase of 12.7% and ORC-2 is associated with a rise of 11.9% in exergy efficiencies for the same increase in isentropic efficiency. The exergetic efficiencies of the solar and geothermal power generation are also observed to increase by 4.1% and 9.1%, respectively.

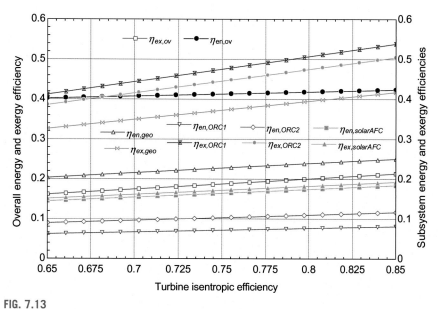

FIG. 7.13

Effect of turbine isentropic efficiency on the efficiencies of the overall system and subsystems.

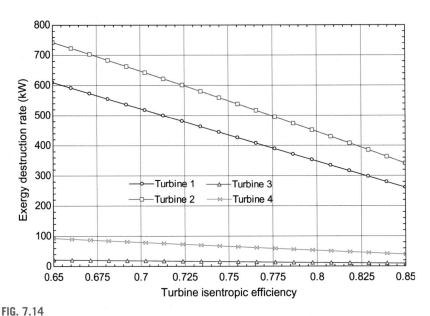

FIG. 7.14

Turbine isentropic efficiency effects on the rates of exergy destruction.

The effects of turbine isentropic efficiency on exergy destruction rates of turbines T-1, T-2, T-3, and T-4 are depicted in Fig. 7.14. The highest reduction in exergy destruction is observed to occur in T-2, where when isentropic efficiency is varied from 65% to 85%, exergy destruction rate reduces by 402.1 kW. Further, turbine T-1 also shows a considerable change in the exergy destruction rates. The rate of exergy destruction reduces from 609.6 to 261.3 kW for the same change in isentropic efficiency. Also, T-3 and T-4 are observed to follow a decreasing trend in exergy destruction rates. The decrease in turbine exergy destruction rates is directly related to the enhancement in both exergetic and energetic performance improvement of the overall system. As the irreversibilities reduce, higher amounts of useful outputs can be obtained resulting in higher efficiencies. Hence, it is recommended to investigate ways of increasing the isentropic efficiencies of turbines as well as other mechanical system components, which can also lead to an improvement in the existing systems.

The effects of geothermal flashing pressure on the overall energetic, as well as exergetic performances, are depicted in Fig. 7.15. The range of flashing pressure is varied from 500 to 2000 kPa and efficiencies are observed to increase up to a maximum point, after which a decreasing trend is observed in the efficiencies. As the flashing pressure increases up to nearly 900–1000 kPa, the efficiencies are observed to rise. However, after this maximum point, the efficiencies start decreasing with increasing flashing pressures. Similarly, the overall exergy efficiency also follows the same trend. This can be attributed to the variation in the vapor quality and specific exergies that results from geothermal fluid flashing. Thus, it is suggested to

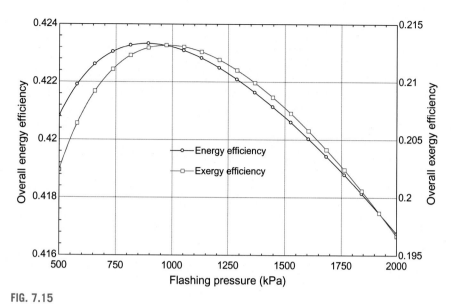

FIG. 7.15

Effect of flashing pressure on the overall energy and exergy efficiencies.

investigate the optimal pressures suitable for flashing providing highest operational energy and exergy efficiencies for any given geothermal-based power plant.

The effects of turbine T-3 inlet pressure on the efficiencies of the overall system as well as ORC 1 are depicted in Fig. 7.16. The overall energy efficiency is found to increase from 42.31% to 42.32% as the inlet pressure varies from 1000 to 2000 kPa. Also, the overall exergetic performance showed a similar trend with a marginal increase from 21.2% to 21.3%. However, the energetic performance of ORC-1 is found to increase from efficiency of 5.3% to 8.8%. In addition to this, the exergetic performance of ORC-1 is observed to rise considerably from 35% to 58.5% as the inlet pressure of T-3 rises from 1000 to 2000 kPa. Thus, during the design phase of a given power plant, it is essential to investigate the operating pressures that provide the optimal performances in terms of both energetic and exergetic efficiency.

Fig. 7.17 depicts the effects of the ambient temperature on energy and exergy efficiencies of the RO subsystem. The exergy efficiency shows a considerable change with the ambient temperature and the energy efficiency is not observed to vary significantly. The RO exergy efficiency, for instance, varies from nearly 35% to 18% as the ambient temperature is increased from 20 to 40°C. This depicts the importance of analyzing the exergy efficiency of a given system. In several cases, energy efficiency solely does not comprehend all the variations in performances with changing operating conditions. As the ambient temperature rises, the specific exergy at each system state point also changes. This leads to a variation in the overall exergetic performances. In integrated energy systems, major system parameters effect the performances and should thus be investigated.

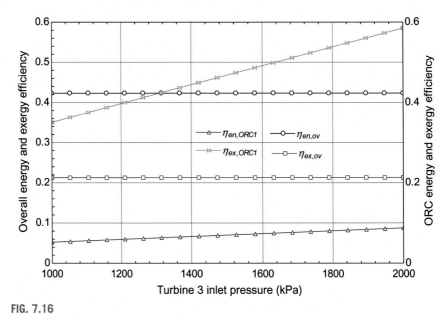

FIG. 7.16

Effect of turbine inlet pressure on the efficiencies of the overall system as well as ORC-1.

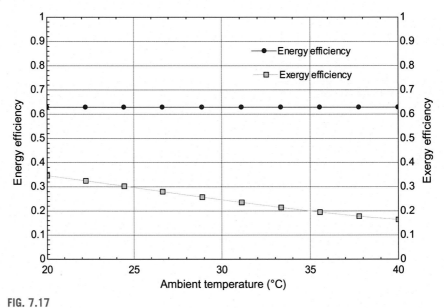

FIG. 7.17

Effect of ambient temperature on the reverse osmosis desalination subsystem efficiencies.

A new solar integrated ammonia fuel cell and geothermal-based energy system are developed for production of power, desalinated water, hydrogen, and cooling. A RO system is used for desalination and a proton exchange membrane electrolyzer is utilized for hydrogen production. Moreover, an absorption cooling system is utilized to provide district cooling through the utilization of system waste heat. Energetic and exergetic performances of the system are studied. The overall energy efficiency of the developed system is found to be 42.3% and the overall exergy efficiency is determined as 21.3%. The integrated multigeneration system results in higher energetic and exergetic efficiencies than geothermal and solar-based power generation subsystems. The major rates of exergy destruction in the system are calculated and the absorption cooling cycle generator, as well as geothermal flash chamber, are found to entail the highest exergy destruction rates of 2370.2 and 643.3 kW. Several parametric analyses are also conducted that provide important insights about the performance of the developed system at varying system parameters.

7.3 Closing remarks

In this chapter, two case studies are presented that discuss new types of ammonia fuel cell-based technologies. The first case study includes a hybrid ammonia fuel cell and battery system with a regenerative nickel electrode, where the regeneration of the electrode is attained through fuel cell operation. Also, the battery operation provides ammonia fuel that is used as input to the fuel cell subsystem. The second case study presents a new integrated solar and geothermal-based energy system incorporating ammonia fuel cells for multigeneration. The renewable energy system provides useful outputs of electricity, cooling, hydrogen, and desalinated fresh water through RO desalination. The comprehensive modeling, analyses, and assessment of each system is discussed to provide insights into the development of new ammonia-based electrochemical systems. The case studies presented provide the detailed thermodynamic analysis equations implemented on each subsystem. Further, the performances of the developed systems are assessed at varying operating conditions and system parameters through different parametric studies. Both the electrochemical as well as overall performances of the developed systems are elucidated with recommendations for further improvements.

Conclusions and future directions

In this book, ammonia fuel cells are elucidated in depth considering all aspects of their development, operation, analysis, assessment, performance, and applications. The underlying concepts and fundamentals are first covered to provide complete understanding about their development and operation. Development of different types of ammonia fuel cells is presented and their performances are comparatively assessed. The modeling of ammonia fuel cells is described in detail considering the determination of various electrochemical parameters to simulate and model a given type of ammonia fuel cell. Furthermore, several new types of ammonia fuel cell-based integrated systems are described and their performances are assessed at varying operating conditions and system parameters.

Ammonia has been used since several decades in refrigeration systems where the ammonia-water absorption cooling cycles were used extensively to produce cooling effect through excess or waste heat. However, as depicted in Fig. 8.1, the evolution of ammonia-based technologies has seen a steady improvement. Ammonia-based boilers are being developed that combust ammonia or ammonia-based fuel blends to provide thermal energy that heats a given fluid. Furthermore, the development of ammonia-based engines is underway wherein first ammonia-blended conventional gasoline and diesel fuels were investigated. This is followed by the development of ammonia-fueled cars that included engines operating with the blends of ammonia and conventional transportation fuels. Furthermore, ammonia gas turbines have been recently introduced in Japan wherein power generated using ammonia and ammonia fuel blends was utilized as energy inputs.

Ammonia fuel cell development has also gained pace in the recent past where new types of ammonia-fueled electrochemical engines have been developed. Fuel cells operating with ammonia fuel at different temperature levels, varying electrolytes, and electrochemical catalysts have been developed. Especially, the commercial development and successful testing of ammonia-fueled solid oxide fuel cells were reported in Japan. Next, the shift toward the new era of energy will entail the development of integrated systems that incorporate several renewable energy resources including ammonia fuel cells to provide several useful commodities. Various types of integrated systems developed in the recent past were presented earlier in this book and further development of integrated systems is underway.

The evolution of ammonia-based energy production into a new era is expected due to the fact that several major companies and corporations have shown interest in the development of new ammonia-based energy technologies. The recent

Ammonia Fuel Cells. https://doi.org/10.1016/B978-0-12-822825-8.00008-6

233

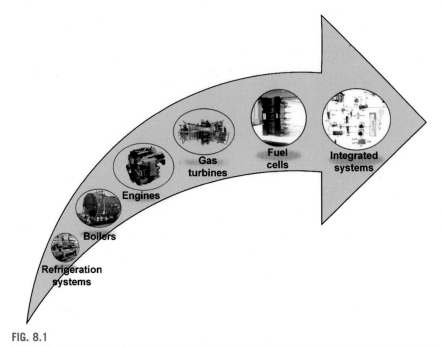

FIG. 8.1

Evolution of ammonia-based technologies.

developments in the area of ammonia energy in Japan are particularly encouraging. A collaboration of various companies and research centers comprising 22 members are working on the development and enhancement of ammonia-based infrastructure. Their objectives include the development of clean hydrogen followed by clean synthesis of ammonia. Further, the utilization of ammonia as a source of energy through ammonia-based combustion engines as well as direct ammonia fuel cells are key objectives of the program. Furthermore, various major companies such as Yara have announced plans toward the development of pilot plants that can produce environmentally benign ammonia. Production of green ammonia will lead to major changes in the present scenario where conventional ammonia synthesis relies heavily on fossil fuel usage. Green ammonia is essential toward the development of direct ammonia fuel cells that are environmentally benign throughout their life cycles. Moreover, Mitsubishi Hitachi Power Systems recently announced their plans toward the usage of ammonia in their power generation technologies. Plans to utilize ammonia along with coal in thermal boilers to reduce greenhouse gas emissions were announced. Moreover, ammonia is becoming an important part of various hydrogen-based strategic plans across the globe. Australia recently introduced its hydrogen action plan in which the development of hydrogen-based infrastructure was announced. Ammonia was described to be one of the major chemicals incorporated in this plan as a medium for hydrogen as well as energy storage. Furthermore, the green ammonia consortium initiated in Japan is also a major advancement in the area of ammonia energy. This collaborative body comprising various multinational corporations as well as research institutes aims to develop a life cycle chain of ammonia that is free of CO_2 emissions. In addition to this, the power to ammonia (P2A) concept is gaining pace and popularity with time where clean synthesis of ammonia from renewable electricity is being developed. Major countries such as China have introduced P2A infrastructural development plans.

8.1 Ammonia engines

Ammonia engines are being developed for different applications in several countries. Several major corporations have announced development projects that are aimed at developing ships that operate with ammonia fuel. In addition, several major companies have announced that various types of maritime transportation options powered with ammonia will be introduced in China in the upcoming years. Also, Japan launched another program aimed at the research and development of maritime engines that operate with carbon-free fuels. Ammonia and hydrogen were identified as the primary candidates for these applications. Moreover, Siemens started an investigation into the usage of ammonia as an energy storage medium where electrical power was utilized to synthesize ammonia and later the produced ammonia was employed for power generation. Such efforts exerted by large commercial corporations show the shift from fossil-based engines to clean engines powered by ammonia or hydrogen, which is to come in the near future. The shift toward this new era of carbon-free energy resources entails both hydrogen and ammonia as key players. The current transportation sector relies heavily on combustion engines. However, the engines in use have been designed for usage with gasoline, diesel, or other hydrocarbon fuels. Hence, the replacement of all these engines across the globe with ammonia powered engines would be a major challenge that needs to be tackled. Some solutions have proposed the development of retrofit ammonia engines that can be used as replacements or modifications to the current gasoline or diesel engines in automobiles. Investigations are underway to introduce such types of technologies that can decrease the usage of gasoline as well as diesel as transportation fuels through the usage of innovative retrofit systems. Since ammonia entails different combustion properties from gasoline and diesel, such as ignition temperatures and pressures, the retrofitting methods are aimed at modifying current spark ignition or compression ignition engines in a way that ammonia can be suitably combusted. Furthermore, the air intake ratio for ammonia-fueled engines is also different from conventional engines and thus air-fuel ratio will have to be modified to suit the usage of ammonia fuel.

8.2 Ammonia gas turbines

Ammonia gas turbines are also being developed in countries such as Japan, United Kingdom, United States, and Turkey aiming at carbon-free power generation. In addition, several research institutions as well as commercial companies have also announced future development plans that aim at producing ammonia-fueled gas turbines. These gas turbines entail the combustion of ammonia fuel as a replacement of natural gas that is utilized extensively as a fuel in numerous gas turbine power plants across the globe. It is often argued whether further development of thermal power plants should be continued owing to the presence of high efficiency fuel cell technologies. Nevertheless, factors such as minimizing energy costs still require the usage of thermal engines. Furthermore, although fuel cells entail higher efficiencies than gas turbine power plants, the combined heat and power plants have been reported to achieve high efficiencies of nearly 50% that still make gas turbines a viable option in terms of operational efficiencies. The usage of combined fuel cells and gas turbines has also been proposed where the integrated usage of high-temperature solid oxide fuel cells as well as gas turbines is undertaken. Moreover, it was reported by Siemens that the development of ammonia combustion-based technologies could open a new era of renewable energy, where renewable and environmentally benign energy could be produced across the globe. Also, the IHI Corporation in Japan has already reported operational trials at an industrial scale for hybrid natural gas and ammonia fired gas turbines. These type of gas turbines include the generation of power through the usage of fuel blends where natural gas is the primary fuel and a portion of ammonia is slowly introduced. As ammonia is introduced along with natural gas, the air-fuel ratio as well as other parameters including combustion temperatures need to be adjusted accordingly to attain suitable power output performance. Nevertheless, the implementation of ammonia gas turbines entails the potential to reduce massive amounts of CO_2 emissions that are currently emitted during the operation of natural gas turbines that are employed in numerous power plants across the globe.

8.3 Ammonia fuel cells

Ammonia fuel cells entail a vital position in the area of carbon-free energy where the usage of ammonia provides several favorable advantages that can solve the current challenges faced by hydrogen fuel cells. Several challenges are also faced by ammonia fuel cells currently that need to be addressed for further development. The primary challenge comprises the scarcity of high compatibility electrochemical catalysts. Platinum black catalyst, which is the most commonly used catalyst in other types of fuel cells, entails high adsorption energy for nitrogen atoms leading to catalyst poisoning during the operation of an ammonia fuel cell. This has been identified as the primary reason for attaining comparatively lower output voltages and power densities than the expected theoretical values. Thus, more ammonia-compatible catalysts need to be developed that can enhance the performance of direct ammonia fuel cells. Such catalysts comprise iron-based composites or alloys that can allow both sufficient electrochemical oxidation of ammonia and lower poisoning of catalysts. Several types of catalysts have been introduced in the recent past to enhance the output voltages of direct ammonia fuel cells. However, fuel cell performances close to the theoretical performance in terms of the open-circuit voltage, peak power density, and short-circuit current density have not yet been attained. Thus, efforts should be directed in this area where both high open-circuit voltage and high peak power densities should be attained with low-temperature direct ammonia fuel cells.

The performance of ammonia-fed solid oxide fuel cells has been comparable to that of hydrogen-fueled cells. This is primarily attributable to the utilization of high operating temperatures that dissociate ammonia before the electrochemical reactions. Nevertheless, given the considerably lower storage costs of ammonia fuel as compared to hydrogen, ammonia can act as an environmentally benign fuel for solid oxide fuel cells. The open-circuit voltages as well peak power densities were also found to be much higher than direct ammonia fuel cells operated at ambient conditions. However, the longevity of these cells with ammonia fuel needs to be further investigated. Some studies have reported insufficient lifetimes of ammonia-fueled solid oxide cells where the performance deterioration with time was observed to be considerable. This was attributed to cell poisoning caused by nitrogen oxide molecules. Methods to overcome these issues need to be developed where ammonia-fueled solid oxide fuel cells can attain longer lifetimes comparable to hydrogen-fueled cells.

References

[1] International Energy Agency (IEA), Global energy and CO_2 status report, Available at: https://www.iea.org/reports/global-energy-and-co2-status-report-2019, 2019.

[2] International Energy Agency (IEA), Data and statistics, Available at: https://www.iea.org/data-and-statistics, 2019.

[3] International Energy Agency (IEA), Total primary energy supply (TPES) by source, Available at: https://www.iea.org/statistics/, 2016.

[4] International Energy Agency (IEA), Global energy & CO_2 status report, Available at: https://www.iea.org/geco/emissions/, 2018.

[5] International Energy Agency (IEA), CO_2 emissions statistics, Available at: https://www.iea.org/statistics/co2emissions/, 2018.

[6] O. Siddiqui, I. Dincer, A review on fuel cell-based locomotive powering options for sustainable transportation, Arab. J. Sci. Eng. 44 (2019) 677–693.

[7] A.J. Appleby, F.R. Foulkes, Fuel Cell Handbook, Van Nostrand Reinhold, New York, 2000.

[8] L. Mond, C. Langer, A new form of gas battery, Proc. R. Soc. London 46 (1889) 296–304.

[9] E. Cairns, E. Simons, A. Tevebaugh, Ammonia–oxygen fuel cell, Nature 217 (1968) 780–781.

[10] Comparison of Fuel Cell Technologies, U.S. Department of Energy, 2019. Available at: http://www.hydrogen.energy.gov.

[11] I. Dincer, C. Acar, Review and evaluation of hydrogen production methods for better sustainability, Int. J. Hydrogen Energy 40 (2015) 11094–11111.

[12] E. Antolini, E.R. Gonzalez, Alkaline direct alcohol fuel cells, J. Power Sources 195 (2010) 3431–3450.

[13] K.S. Roelofs, T. Hirth, T. Schiestel, Dihydrogenimidazole modified silica-sulfonated poly (ether ether ketone) hybrid materials as electrolyte membranes for direct ethanol fuel cells, Mater. Sci. Eng. B 176 (2011) 727–735.

[14] G. Andreadis, V. Stergiopoulos, S. SOng, P. Tsiakaras, Direct ethanol fuel cells: the effect of the cell discharge current on the products distribution, Appl. Catal. Environ. 100 (2010) 157–164.

[15] L. Jiang, G. Sun, S. Wang, G. Wang, Q. Xin, Z. Zhou, Electrode catalysts behaviour during direct ethanol fuel cell life-time test, Electrochem. Commun. 7 (2005) 663–668.

[16] Renewable Fuels Association. "Renewable Fuels Association Industry Statistics". Renewable Fuels Association. 2019. Available from: https://ethanolrfa.org/statistics/. Retrieved 2017-04-23.

[17] L. Gong, Z. Yang, K. Li, W. Xing, C. Liu, J. Ge, Recent development of methanol electrooxidation catalysts for direct methanol fuel cell, J. Energy Chem. 27 (2018) 1618–1628.

[18] D. Cao, S.H. Bergens, A direct 2-propanol polymer electrolyte fuel cell, J. Power Sources 124 (2003) 12–17.

[19] I. Chino, K. Hendrix, A. Keramati, O. Muneeb, J.L. Haan, A split pH direct liquid fuel cell powered by propanol or glycerol, Appl. Energy 251 (2019) 113323.

[20] M.E.P. Markiewicz, S.H. Bergens, A liquid electrolyte alkaline direct 2-propanol fuel cell, J. Power Sources 195 (2010) 7196–7201.

[21] Bloomenergy, 2019. Available at: http://www.bloomenergy.com/clean-energy. (Accessed 7 December 2019).

[22] S. Liu, Y. Behnamian, K.T. Chuang, Q. Liu, J. Luo, A-site deficient $La_{0.2}Sr_{0.7}TiO_{3-\delta}$ anode material for proton conducting ethane fuel cell to cogenerate ethylene and electricity, J. Power Sources 298 (2015) 23–29.

[23] J. Li, X. Fu, J. Luo, K.T. Chuang, A.R. Sanger, Evaluation of molybdenum carbide as anode catalyst for proton-conducting hydrogen and ethane solid oxide fuel cells, Electrochem. Commun. 15 (2012) 81–84.

[24] X. Fu, X. Luo, J. Luo, K.T. Chuang, A.R. Sanger, A. Krzywicki, Ethane dehydrogenation over nano-Cr_2O_3 anode catalyst in proton ceramic fuel cell reactors to co-produce ethylene and electricity, J. Power Sources 196 (2011) 1036–1041.

[25] X. Fu, J. Luo, A.R. Sanger, N. Luo, K.T. Chuang, Y-doped $BaCeO_3 - \delta$ nanopowders as proton-conducting electrolyte materials for ethane fuel cells to co-generate ethylene and electricity, J. Power Sources 195 (2010) 2659–2663.

[26] Z. He, C. Li, C. Chen, Y. Tong, T. Luo, Z. Zhan, Membrane-assisted propane partial oxidation for solid oxide fuel cell applications, J. Power Sources 392 (2018) 200–205.

[27] J. Hong, M. Usman, S. Lee, R. Song, T. Lim, Thermally self-sustaining operation of tubular solid oxide fuel cells integrated with a hybrid partial oxidation reformer using propane, Energ. Conver. Manage. 189 (2019) 132–142.

[28] Field test of propane fuel cell in Alaska, Fuel Cells Bull. 2007 (2007) 3–4.

[29] Y. Feng, J. Luo, K.T. Chuang, Carbon deposition during propane dehydrogenation in a fuel cell, J. Power Sources 167 (2007) 486–490.

[30] K. Wang, P. Zeng, J. Ahn, High performance direct flame fuel cell using a propane flame, Proc. Combust. Inst. 33 (2011) 3431–3437.

[31] T. Huang, C. Wu, C. Wang, Fuel processing in direct propane solid oxide fuel cell and carbon dioxide reforming of propane over Ni–YSZ, Fuel Process. Technol. 92 (2011) 1611–1616.

[32] M.T. Mehran, S. Park, J. Kim, J. Hong, S. Lee, S. Park, R. Song, J. Shim, T. Lim, Performance characteristics of a robust and compact propane-fueled 150 W-class SOFC power-generation system, Int. J. Hydrogen Energy 44 (2019) 6160–6171.

[33] Y. Zhang, F. Yu, X. Wang, Q. Zhou, J. Liu, M. Liu, Direct operation of Ag-based anode solid oxide fuel cells on propane, J. Power Sources 366 (2017) 56–64.

[34] C. Thieu, H. Ji, H. Kim, K.J. Yoon, J. Lee, J. Son, Palladium incorporation at the anode of thin-film solid oxide fuel cells and its effect on direct utilization of butane fuel at 600 °C, Appl. Energy 243 (2019) 155–164.

[35] K. Hayashi, O. Yamamoto, H. Minoura, Portable solid oxide fuel cells using butane gas as fuel, Solid State Ion. 132 (2000) 343–345.

[36] H. Sumi, T. Yamaguchi, K. Hamamoto, T. Suzuki, Y. Fujishiro, Impact of direct butane microtubular solid oxide fuel cells, J. Power Sources 220 (2012) 74–78.

[37] G. Kaur, S. Basu, Performance studies of copper–iron/ceria–yttria stabilized zirconia anode for electro-oxidation of butane in solid oxide fuel cells, J. Power Sources 241 (2013) 783–790.

[38] J. Jeong, S. Baek, J. Bae, A diesel-driven, metal-based solid oxide fuel cell, J. Power Sources 250 (2014) 98–104.

[39] T.A. Trabold, J.S. Lylak, M.R. Walluk, J.F. Lin, D.R. Troiani, Measurement and analysis of carbon formation during diesel reforming for solid oxide fuel cells, Int. J. Hydrogen Energy 37 (2012) 5190–5201.

[40] M.R. Walluk, J. Lin, M.G. Waller, D.F. Smith, T.A. Trabold, Diesel auto-thermal reforming for solid oxide fuel cell systems: anode off-gas recycle simulation, Appl. Energy 130 (2014) 94–102.

[41] K. Zhao, X. Hou, M.G. Norton, S. Ha, Application of a NiMo–$Ce_{0.5}Zr_{0.5}O_{2-\delta}$ catalyst for solid oxide fuel cells running on gasoline, J. Power Sources 435 (2019) 226732.

[42] X. Hou, O. Marin-Flores, B.W. Kwon, J. Kim, M.G. Norton, S. Ha, Gasoline-fueled solid oxide fuel cell using MoO_2-based anode, J. Power Sources 268 (2014) 546–549.

[43] K. Zhao, X. Hou, Q. Bkour, M.G. Norton, S. Ha, NiMo-ceria-zirconia catalytic reforming layer for solid oxide fuel cells running on a gasoline surrogate, Appl. Catal. Environ. 224 (2018) 500–507.

[44] Y. Bicer, I. Dincer, G. Vezina, F. Raso, Impact assessment and environmental evaluation of various ammonia production processes, Environ. Manag. 59 (2017) 842–855.

[45] Ammonia | Industrial Efficiency Technology Database; Measures, 2014. Available from: http://www.iipinetwork.org/wp-content/Ietd/content/ammonia.html#technology-resources.

[46] International Energy Agency (IEA), Energy Technology Perspectives, OECD Publishing, France, 2012. https://doi.org/10.1787/energy_tech-2012-en.

[47] Q. Ma, R. Peng, L. Tian, G. Meng, Direct utilization of ammonia in intermediate-temperature solid oxide fuel cells, Electrochem. Commun. 8 (2006) 1791–1795.

[48] G. Meng, C. Jiang, J. Ma, Q. Ma, X. Liu, Comparative study on the performance of a SDC-based SOFC fueled by ammonia and hydrogen, J. Power Sources 173 (2007) 189–193.

[49] M. Liu, R. Peng, D. Dong, J. Gao, X. Liu, G. Meng, Direct liquid methanol-fueled solid oxide fuel cell, J. Power Sources 185 (2008) 188–192.

[50] Z. Limin, C. You, Y. Weishen, L.I.N. Liwu, A direct ammonia tubular solid oxide fuel cell, Chin. J. Catal. 28 (2007) 749–751.

[51] Q. Ma, J. Ma, S. Zhou, R. Yan, J. Gao, G. Meng, A high-performance ammonia-fueled SOFC based on a YSZ thin-film electrolyte, J. Power Sources 164 (2007) 86–89.

[52] G.G.M. Fournier, I.W. Cumming, K. Hellgardt, High performance direct ammonia solid oxide fuel cell, J. Power Sources 162 (2006) 198–206.

[53] A. Fuerte, R.X. Valenzuela, M.J. Escudero, L. Daza, Ammonia as efficient fuel for SOFC, J. Power Sources 192 (2009) 170–174.

[54] L. Pelletier, A. McFarlan, N. Maffei, Ammonia fuel cell using doped barium cerate proton conducting solid electrolytes, J. Power Sources 145 (2005) 262–265.

[55] N. Maffei, L. Pelletier, J.P. Charland, A. McFarlan, An intermediate temperature direct ammonia fuel cell using a proton conducting electrolyte, J. Power Sources 140 (2005) 264–267.

[56] N. Maffei, L. Pelletier, J.P. Charland, A. McFarlan, An ammonia fuel cell using a mixed ionic and electronic conducting electrolyte, J. Power Sources 162 (2006) 165–167.

[57] N. Maffei, L. Pelletier, A. McFarlan, A high performance direct ammonia fuel cell using a mixed ionic and electronic conducting anode, J. Power Sources 175 (2008) 221–225.

[58] Q. Ma, R. Peng, Y. Lin, J. Gao, G. Meng, A high-performance ammonia-fueled solid oxide fuel cell, J. Power Sources 161 (2006) 95–98.

[59] L. Zhang, W. Yang, Direct ammonia solid oxide fuel cell based on thin proton-conducting electrolyte, J. Power Sources 179 (2008) 92–95.

[60] Y. Lin, R. Ran, Y. Guo, W. Zhou, R. Cai, J. Wang, Z. Shao, Proton-conducting fuel cells operating on hydrogen, ammonia and hydrazine at intermediate temperatures, Int. J. Hydrogen Energy 35 (2010) 2637–2642.

[61] K. Xie, Q. Ma, B. Lin, Y. Jiang, J. Gao, X. Liu, G. Meng, An ammonia fuelled SOFC with a $BaCe_{0.9}Nd_{0.1}O_3$ thin electrolyte prepared with a suspension spray, J. Power Sources 170 (2007) 38–41.

[62] J.C. Ganley, An intermediate-temperature direct ammonia fuel cell with a molten alkaline hydroxide electrolyte, J. Power Sources 178 (2008) 44–47.

[63] J. Yang, H. Muroyama, T. Matsui, K. Eguchi, Development of a direct ammonia-fueled molten hydroxide fuel cell, J. Power Sources 245 (2014) 277–282.

[64] O. Siddiqui, I. Dincer, Experimental investigation and assessment of direct ammonia fuel cells utilizing alkaline molten and solid electrolytes, Energy 169 (2019) 914–923.

[65] R. Lan, S. Tao, Direct ammonia alkaline anion-exchange membrane fuel cells, Electrochem. Solid St. 13 (2010) B83.

[66] S. Suzuki, H. Muroyama, T. Matsui, K. Eguchi, Fundamental studies on direct ammonia fuel cell employing anion exchange membrane, J. Power Sources 208 (2012) 257–262.

[67] P. Olu, F. Deschamps, G. Caldarella, M. Chatenet, N. Job, Investigation of platinum and palladium as potential anodic catalysts for direct borohydride and ammonia borane fuel cells, J. Power Sources 297 (2015) 492–503.

[68] X. Zhang, S. Han, J. Yan, M. Chandra, H. Shioyama, K. Yasuda, N. Kuriyama, T. Kobayashi, Q. Xu, A new fuel cell using aqueous ammonia-borane as the fuel, J. Power Sources 168 (2007) 167–171.

[69] X. Zhang, J. Yan, S. Han, H. Shioyama, K. Yasuda, N. Kuriyama, Q. Xu, A high performance anion exchange membrane-type ammonia borane fuel cell, J. Power Sources 182 (2008) 515–519.

[70] H.G. Oswin, M. Salomon, The anodic oxidation of ammonia at platinum black electrodes in aqueous KOH electrolyte, Can. J. Chem. 41 (1963) 1686–1694.

[71] O. Siddiqui, I. Dincer, Analysis and performance assessment of a new solar-based multigeneration system integrated with ammonia fuel cell and solid oxide fuel cell-gas turbine combined cycle, J. Power Sources 370 (2017) 138–154.

[72] O. Siddiqui, I. Dincer, Development and assessment of a novel integrated system using an ammonia internal combustion engine and fuel cells for cogeneration purposes, Energy Fuel 33 (2019) 2413–2425.

[73] O. Siddiqui, I. Dincer, Design and analysis of a novel solar-wind based integrated energy system utilizing ammonia for energy storage, Energ. Conver. Manage. 195 (2019) 866–884.

[74] X. Li, Principles of Fuel Cells, Taylor & Francis, New York, 2006.

[75] M.M. Mench, Fuel Cell Engines, John Wiley & Sons, New York, 2008.

[76] D.P. Bhatt, R. Udhayan, Electrochemical studies on a zinc-lead-cadmium alloy in aqueous ammonium chloride solution, J. Power Sources 47 (1994) 177–184.

[77] H. Sato, M. Yui, H. Yoshikawa, Ionic diffusion coefficients of Cs^+, Pb^{2+}, Sm^{3+}, Ni^{2+}, SeO^{2-}_4 and TcO^-_4 in free water determined from conductivity measurements, J. Nucl. Sci. Technol. 33 (1996) 950–955.

[78] E. Samson, J. Marchand, K.A. Snyder, Calculation of ionic diffusion coefficients on the basis of migration test results, Mater. Struct. 36 (2003) 156–165.

[79] J.C. Ganley, Ammonia Fuel Cell Systems, Howard University, Department of Chemical Engineering, 2017.

[80] C. Zamfirescu, I. Dincer, Thermodynamic analysis of a novel ammonia–water trilateral Rankine cycle, Thermochim. Acta 477 (2008) 7–15.

Index

Note: Page numbers followed by *f* indicate figures and *t* indicate tables.

Printed in the United States
By Bookmasters